Dynamical Meteorology

AN INTRODUCTORY SELECTION

Edited by

B. W. Atkinson

Queen Mary College,
University of London

Methuen

London and New York

First published in 1981 by
Methuen & Co. Ltd
11 New Fetter Lane, London EC4P 4EE
Published in the USA by
Methuen & Co.
in association with Methuen, Inc.
733 Third Avenue, New York, NY 10017

Printed in the United States of America

British Library Cataloguing in Publication Data
Dynamical meteorology.
1. Meteorology
I. Atkinson, Bruce Wilson
551.5   QC861.2   80-41675

ISBN 0-416-73830-3
ISBN 0-416-73840-0 (University paperback 738)

Contents

List of contributors

B. W. Atkinson Department of Geography, Queen Mary College, University of London.

A. J. Gadd Dynamical Climatology Branch, Meteorological Office, Bracknell.

J. S. A. Green Atmospheric Physics Group, Department of Physics, Imperial College, University of London.

R. S. Harwood Department of Meteorology, University of Edinburgh.

A. Ibbetson Department of Meteorology, University of Reading.

H. A. Panofsky Department of Meteorology, Pennsylvania State University.

M. A. Pedder Department of Meteorology, University of Reading.

E. R. Reiter Department of Atmospheric Science, Colorado State University.

J. Smagorinsky Geophysical Fluid Dynamics Laboratory, Princeton University.

A. A. White Geophysical Fluid Dynamics Laboratory, Meteorological Office, Bracknell.

Preface

This volume was produced by 'popular demand'. The bulk of its contents first appeared as a series of articles in *Weather*, the monthly magazine of the Royal Meteorological Society, between January 1978 and April 1979. The response to the series was positive, including several requests that the articles be collected into one volume. This idea met with the approval of the Council of the Royal Meteorological Society and had, independently, attracted Methuen Publishers. The present collection results from following through these initiatives.

Over half the readership of *Weather* have no professional qualification in meteorology but do have a very strong interest in the ways of the atmosphere. The increasing complexity of meteorology means that not only the non-professional but also many professional meteorologists find the subject difficult to understand. The original aim of the series was therefore to try to simplify some of the more difficult concepts in dynamical meteorology, the heart of theoretical meteorology. In the first chapter of this collection I try to show why physics and mathematics have any relevance at all to the weather and climate we all experience. The remainder of the collection attempts to elucidate their application to the atmosphere.

At least three problems confront a compiler of such a series. First, is the problem of balance of contents. The material in Chapters 1–8 will be comparatively stock-in-trade to the professional meteorologist. But it is precisely this material that the non-professional, including the complete layman and many professionals in other disciplines, such as geography and engineering, may find useful. Even professional meteorologists have said that they found this material of value on its first appearance. On the other hand, the non-professional may find the material in Chapters 9–15 rather hard going the first time around. This material could prove of more immediate interest to the meteorologist with a firm grounding in the elementary theory. Nevertheless, with perseverance the non-professional could learn much from Chapters 9–15.

The second major problem lies in the use of mathematics, particularly calculus. Many non-meteorologists with an interest in the atmosphere, such as geographers,

do not have a good knowledge of calculus. Because much of dynamical meteorology is expressed in the form of differential equations this lack is a major
stumbling block to communication. As noted in Chapter 1, it is impossible to
give a crash course in calculus in such a collection, but it is possible for the non-
mathematician to get a grasp of the main theoretical ideas once the psychological
barrier of mathematical symbols has been overcome. Perhaps the most important
symbols to understand are d/dt and $\partial/\partial t$. The former means the rate of change
with time of the property on which d/dt operates (say x) as the particle with
that property moves through *space* and *time*. Thus if x is a wind-velocity (say u),
then du/dt means that change in u as the air particle moves: if one could sit on
the particle du/dt would measure the change in velocity. The evaluation of this
derivative, as it is called, would be achieved by 'following the motion' to use the
phraseology of the meteorological literature. In contrast, the symbols $\partial/\partial t$ mean
the change in a property *at a fixed point*, and clearly the derivative is *not* evaluated
by following the motion. Of these two ways of expressing changes with time the
first is known as the Lagrangian way and the second as the Eulerian. The two
sets of symbols in an atmospheric context with wind components u (zonal — x),
v (meridional — y), w (vertical — z) are related as follows:

$$\frac{d}{dt} = \frac{\partial}{\partial t} + u\frac{\partial}{\partial x} + v\frac{\partial}{\partial y} + w\frac{\partial}{\partial z}.$$

In this expression, the symbols $\partial/\partial x$, $\partial/\partial y$, $\partial/\partial z$ mean rate of change with,
respectively, x, y, and z. Even with this limited knowledge the non-mathematician
may find some of the equations in this volume much less intimidating. Of course,
the reader with a good knowledge of elementary calculus will find the going far
easier.

The third problem is that of deciding what should or should not be included
in the series. Clearly, I could not, nor did I desire to be, comprehensive. Equally
clearly, my choice of topics will not satisfy everyone. Perhaps the basic reason
for inclusion in the present volume is that, in my view, these particular topics
are widely considered to be of basic importance in dynamical meteorology and,
at the same time, are precisely the subjects that are least well-known by non-
professionals. If there are other basic topics that should have been covered,
perhaps a second volume is called for.

In preparing the original series for the collection I have taken the opportunity
to edit the initial articles into what I hope is a coherent volume. In addition, I
have added five more chapters covering material which was unavailable for the
original series. Three of these chapters (by M. A. Pedder) consider some of the
practical facets of dynamical meteorology, thus complementing the theoretical
material in Chapters 2—4. In Chapter 8, we take the theory a little further to
discover why waves, in all their various forms, seem to occupy such a prominent
place within dynamical meteorology. Chapter 9 is an opportunity 'to pause for
thought', giving a brief summary of progress in dynamical meteorology over the
century 1850—1950. To those unfamiliar with the story, it may provide a
context (and possibly whet an appetite for more personal historical forays),

which itself helps an understanding of the material in Chapters 10–15. The slight overlap of material in Chapters 9 and 15 serves only to stress the importance of the scientific discoveries reported therein.

In conclusion, I should stress that this is not a text book in dynamical meteorology in the accepted sense of the word. It is an attempt to present some complicated facets of the subject in a way simpler than would be found in such standard texts. Those who are disatisfied with such 'easy meat' should turn to the rigours of the many first-class texts now available. Finally, I thank the contributors to the series. They all laboured hard on pieces which could easily have been seen as chores. More than once I heard the cry: "I find it so difficult to simplify things!" Perhaps none of us completely succeeded in our efforts, but the pieces are presented in the hope that readers will begin to feel the excitement of the understanding of and insight into atmospheric behaviour that emerges from a study of dynamical meteorology.

B. W. Atkinson

1
Weather, meteorology, physics, mathematics

B. W. ATKINSON
Queen Mary College,
University of London

We all know what *meteorology* is about — or at least we think we do. Most of us would, in one way or another, say that it is essentially about 'the *weather* . . . and things'. Yet if the layman were to catch a glimpse of the pages of the *Quarterly Journal of the Royal Meteorological Society* or the *Journal of Atmospheric Sciences* he would no doubt wonder what on earth the hieroglyphics therein could remotely have to do with the weather — or the things. Over the last century, and particularly within the last four decades, meteorology appears to have been transformed from the interested and keen observation of sunshine, cloud, rain, snow, temperature, humidity and wind etc. to an incomprehensible mass of abstractions which appear to have little connection with what is going on outside.

Our man, or woman, in the street would not be alone in his or her confusion: at least four eminent meteorologists, while not themselves being confused, have recognized the problem. In his Presidential Address to the Royal Meteorological Society in 1944, Professor D. Brunt (1944, p. 5) declared that the dynamical aspect (involving *mathematical* and *physical* principles) was 'by far the most difficult aspect of meteorology' and appealed for more meetings whose specific aim was to make fairly difficult ideas and techniques 'understandable to all' (p. 11). Little progress in communication appears to have been made along these lines as, 19 years later, Dr H. L. Penman (1963) was still lamenting the lack of communication and, in considering his equation representing changes in meteorology, stressed the 'imaginary term representing the contribution of mathematics'. Difficult or not, the applied mathematicians have been busy in the last 30 years so that their studies have now truly 'displaced description in our literature' (Stagg, 1960, p. 295). The problem is well summarized by the following words of Sir Graham Sutton (1954): 'The basis of pure mathematics, and therefore of all mathematics, is the concept of number, which itself is a creation of human intellect. The introduction of number into logical reasoning has produced mathematics, which thus deals entirely with abstractions and has no essential connection with the world of concrete subjects which we know through our senses. *There is no a priori reason why we should expect rainfall, for example, to*

1

have anything in common with this system of thought (my italics) and to our not very remote ancestors (and even, perhaps, to some men of our own time) it would seem unreasonable . . . to attempt to calculate in advance the amount of rain which falls on the growing crop.'

Despite these eminent reservations it is highly unlikely that today hundreds of highly qualified mathematicians and physicists around the world are busily solving their equations (or trying to) without some 'respectable' meteorological aim in mind. This being the case this chapter outlines some of the different approaches to studying the weather, showing the links between abstraction and rainfall. It acts as an introduction to a series of chapters which attempts to outline some of the important elements of both well-known classical dynamics within the atmosphere and more recent and complicated developments within dynamical meteorology. To many readers, the more elementary chapters will present very familiar material: to others, possibly in the majority, the material will probably be new but, we hope, not incomprehensible and uninteresting. One or two of the chapters will probably make some professional meteorologists scratch their heads at the use of well-known tools to fashion new insights into atmospheric dynamics.

Historical perspective

Most stories benefit from some historical perspective and this one is no exception. Indeed it is only by tracing through time our different approaches to the study of the weather that we can identify the significant links. It is possible, of course, to go as far back as the Greeks but it is more profitable to restrict our period to the last 350 years. By doing so, we avoid the intellectual Dark Ages prior to 1600 and yet retain the important early stages of modern science. In this account we make brief mention of both *observation* and *theory*.

Observation

Early observations of atmospheric characteristics were, of course, made by the human senses of sight, hearing, smell and touch. Whilst 'Dark Age Man' must have been painfully aware of the harshness of some of the weather elements, many centuries passed before it was realized that air had mass and weight and that instruments could be made to measure that weight. In retrospect, this realization was quite stunning, as were many of those of the seventeenth century. The invention of the thermometer and barometer in the mid-seventeenth century meant that two important descriptive elements of the atmosphere could be observed quantitatively for the first time. Instruments to measure humidity, wind speed and direction and precipitation were developed spasmodically: the simpler ones, such as the rain-gauge, being used in Britain early in the eighteenth century; the more complicated ones, such as reliable, accurate anemographs being developed in the late nineteenth century.

The simplicity of the thermometer and rain-gauge meant that gentlefolk with time and money, but not necessarily much instrumental expertise, could observe the 'weather' as frequently as they pleased. As Manley (1974) has clearly demonstrated we owe much to the enthusiasm of these early observers in our studies of climatic change. This tradition of careful observation of the weather remains in Britain, perhaps best exemplified by the fact that over 7000 individuals observe daily rainfall amounts for the Meteorological Office.

Theory

Notwithstanding the significance of observation to the development of science, theory lies at its heart. In the simplest terms, once we know how wet, dry, warm, cold, calm or windy a place is, we want to know why. Another look at the elements most frequently observed gives us a hint of one possible reason. Cloud results from air rising, cooling and water vapour condensing. Precipitation results from interactions between micro-scale processes involving droplets and ice crystals and updraughts within the cloud. Humidity results partially from evaporation at the earth's surface of water and its upward transfer into the atmosphere. Sunshine occurs at the surface in the absence of cloud. Even temperature changes at one place may be a partial result of air masses arriving from elsewhere. In all these cases, the common factor is moving air, and the other frequently observed element, wind, *is* of course air moving relative to the underlying earth. It would appear that if we can explain how air moves, we shall have made significant steps towards explaining 'weather'. In fact, as we shall see, analysis of *air motion* is the central theme of dynamical meteorology.

Once our attention has been focused on air motion, it soon becomes clear that it is a very complicated phenomenon including, at one extreme, small-scale features visible in the movements of cigarette smoke to, on the other extreme, the massive circulations spectacularly shown on speeded-up films taken by geostationary satellites above the earth. But this appreciation of scale is in itself an important step towards understanding air motion. We shall see later in the volume how powerful is the scale concept and how frequently it is used in meteorology.

Dynamics

Setting aside for the moment the problem of scale, once the air movement, so critical to the development of weather, has been described, it must itself be explained. It is possible to apply to the problems of air motion the principles of classical mechanics. This is usually the point where the non-mathematical reader switches off but, in fact, when these principles are conveyed in words as well as mathematical symbols, they are seen to be comparatively painless. The next two chapters by Panofsky (1981a, b) attempt this exercise in communication. At present we lay some foundations.

Important as were the observational advances of the seventeenth century, the theoretical development of mechanics, largely by Sir Isaac Newton, overshadowed them. Among his many contributions to science, Newton's formulation of his three laws of motion and his 'invention' of calculus are the foundations of, not only dynamical meteorology, but also many other aspects of physical science. For those readers not familiar with Newton's Laws of motion, they are as follows:

1. Every body continues in its state of rest or uniform motion in a straight line unless it is acted upon by a force.
2. When a body is accelerated, the magnitude of the force causing the acceleration is equal to the product of the mass of the body and the magnitude of the acceleration. The direction of the force is the same as that of the acceleration. In symbols, the first sentence becomes:

$$F = ma,$$ (1.1)

where F is the force,

m is the mass of the body,
a is the acceleration.

3. To every action there is an equal and opposite reaction.

Of the three laws, the second one is the least 'obvious' and provides the real basis of dynamical meteorology. Force is not an easy concept to put over rigorously in a few words, but its essence is typified by our experience that if we push a pencil, we exert a force and the pencil moves. Our common experience of acceleration is usually that of a vehicle starting from rest, frequently moving in a straight line – i.e. it is the change of speed with time. In fact, acceleration also occurs when the vehicle changes direction. So if a car is driven around in a circle at a constant *speed*, it is still being accelerated because it is changing its direction. The term 'velocity' is used to describe the speed and direction of a body, so acceleration is a change of velocity.

The next step is to show how equation 1.1 has any connection with our weather. We have shown earlier that much of our weather results directly or indirectly from moving air. Consequently, because air has mass and moves, there is a reasonable chance that weather might be related to equation 1.1. The two remaining requirements are a force and the certainty that not only is the air moving, but that it is accelerating. In fact, both are readily to hand within the atmosphere. We are all familiar with the pressure gradient on the weather map. It is due to the uneven distribution of the mass of air in the atmosphere and provides a force which tries to move air from high to low pressure in an attempt to eradicate the uneven mass distribution. The details of this force are outlined by Panofsky (1981a). We now have both a mass of air and a force acting upon it. According to Newton's second law, it must be accelerating whether or not it is immediately obvious to the observer. This is one of the main stumbling blocks in putting over the concepts of dynamical meteorology. We are all familiar with wind speed and we have instruments to measure it regularly. Yet such experience does little to foster the appreciation that even a steady wind comprises air which is accelerating, because it is on a rotating planet and thus is constantly changing its direction, rather like the car being driven around in circles.

It makes sense then to apply Newton's second law to atmospheric motion. Unfortunately for many of us, its application usually involves the use of one of Newton's other major contributions – calculus. It is clearly impossible to give a crash course in calculus in this article, but it may help readers to explain some symbols. Very simply, the collection of symbols d/dt means 'the rate of change with t of' whatever this symbol prefixes. Thus dx/dt means the rate of change of x with t. If x were distance and t were time, dx/dt would mean the speed. The x and t could be other letters of the alphabet and could represent virtually anything that makes sense mechanically (e.g. mass, volume, heat, distance, etc.). Thus, if we call a velocity 'u', then du/dt means a change of velocity with time, that is an acceleration. If we call pressure p, then dp/dx means change of pressure with x which could be distance. This represents a pressure gradient which would cause a force to act on the air within it. We can thus re-write Newton's second law in the following symbolic form:

$$-\frac{1}{\rho}\frac{dp}{dx} = \frac{du}{dt},$$

Force   Acceleration of air

where ρ is the density (i.e. mass per unit volume) of the air.

Due to the universality of Newton's Laws we can apply his second law to any particular bit of the atmosphere we wish. In fact, we can apply it to a multitude of points, as shown by Gadd (1981) later in this volume. Taken to its limits, we can apply it to every point in the atmosphere and, this being the case, the second law clearly describes the behaviour of air motion in a very fundamental way.

We noted earlier that it was far from obvious that the winds were accelerating. We now know that they are and why they are doing so, but, in itself, this knowledge does not help us to discover the wind velocity at any place at any time. In our car analogue, we may know that it is *accelerating* down a particular piece of motorway at a particular time due to the driver 'putting his foot down', but that does not tell us his *speed* – and his speed may be important for several reasons, e.g. petrol consumption or being stopped for speeding. Similarly, in meteorology, we wish to know air velocity because of its paramount importance in the genesis of much of our weather. But we can simply observe that, one may retort, so why bother with all this physics and mathematics? Certainly we can observe it now, but that would not help us to predict what it would be like in the future. With the aid of some rather complicated mathematics it is now possible to derive the wind speed from Newton's second law when it is applied to the atmosphere. This is sometimes called 'solving' the equation, a term much used in subsequent chapters. Consequently, we have a powerful tool for *prediction* and most scientists believe that consistently good predictions indicate a good measure of *understanding* of the system being analysed.

Gas laws

Up to now we have simply noted that air has mass with no further elaboration. Experiments in the eighteenth and nineteenth centuries established that air comprises a mixture of gases, mainly nitrogen and oxygen. With little error, this mixture may be considered to be a 'perfect gas' and consequently behave according to the gas laws which, once again, were discovered in the seventeenth century. These laws relate the pressure, density and temperature of gases in an unambiguous way and, consequently, are powerful tools in the elucidation of atmospheric behaviour. In particular, they are helpful in the analysis of atmospheric thermodynamics.

Thermodynamics

If the reader blanched at the sight of 'dynamics', 'thermodynamics' is likely to cause more serious physiological responses. Yet the aims, if not the methods, of thermodynamics are comparatively simple. They are to analyse the ways in which changes in heat content (using these words in a general sense) of a substance (usually a fluid) affect the dynamics of the substance. A typical thermodynamic process involves the addition of heat to a fluid causing its pressure and volume to change. The laws of thermodynamics relate these changes in heat content, pressure and volume. Clearly the atmosphere, considered to be a perfect gas, is an attractive area for application of these laws. It has mass and volume, exerts a pressure and both receives heat from the sun and loses it to space.

We can now tie together these strands of physical science to show their relevance to meteorology. As just noted, if the sun shines on the atmosphere, the gases change their volume and pressure, and thus pressure gradients are changed. If the atmosphere were initially at rest (a state which can be created in mathematical models) pressure gradients would be created by uneven heating of the atmosphere. Once we have the pressure gradients we induce air to accelerate and this air motion, in a myriad of shapes and sizes and in association with water vapour, results in our wind, cloud, rain and snow.

Conclusion

This introductory chapter has attempted to show how *weather* and *mathematics* are linked through the application of physical ideas to the atmosphere. It has purposely been kept brief and very simple. The aim has *not* been to present a lot of mathematics describing the well-known physical laws. Rather it has been to show why those physical laws, whether they be written in words or symbols, are of any relevance at all to the study of the weather. Once this relevance is appreciated, the next step is to gain an understanding of the physics and this involves the use of mathematics. The following two chapters attempt to take this step in as painless a way as could be devised by the author. In turn they are followed by others which require a deeper understanding: as such they may be beyond the non-professional but, hopefully, will be of interest to professionals not primarily concerned with dynamical meteorology.

References

Brunt, D. (1944) 'Progress in meteorology', *Quart. J. R. Met. Soc.*, 70, 1–11.

Gadd, A. J. (1981) 'Numerical modelling of the atmosphere', this volume, 194–204.

Manley, G. (1974) 'Central England temperatures: monthly means 1659 to 1973', *Quart. J. R. Met. Soc.*, 100, 389–405.

Panofsky, H. A. (1981a) 'Atmospheric hydrodynamics', this volume, 8–20.

Panofsky, H. A. (1981b) 'Atmospheric thermodynamics', this volume, 21–32.

Penman, H. I. (1963) 'Mirror for meteorologists', *Quart. J. R. Met. Soc.*, 89, 453–60.

Stagg, J. M. (1960) 'Has meteorology become too professionalized?' *Quart. J. R. Met. Soc.*, 86, 295–300.

Sutton, O. G. (1954) 'The development of meteorology as an exact science', *Quart. J. R. Met. Soc.*, 80, 328–38.

2
Atmospheric hydrodynamics

H. A. PANOFSKY
Pennsylvania State University

In Chapter 1 Atkinson (1981) showed that *air motion* is fundamental to weather and climate. He also showed that our understanding of air motion relies on the application to the atmosphere of some concepts from classical physics — particularly those from the subjects *hydrodynamics* and *thermodynamics*. The state of the atmosphere is well specified when the distribution of seven characteristics (or variables as they are sometimes called) is known throughout: pressure, temperature, air density, amount of water vapour, wind speed, wind direction and vertical wind. The behaviour of the seven variables is governed by seven equations, three of which are derived from the field of thermodynamics and four from the field of hydrodynamics. The thermodynamic equations include the gas law, a heat equation and an equation expressing conservation of moisture. The hydrodynamic equations are the three equations of motion (components of Newton's second law (see Atkinson, 1981) in three directions) and the equation of continuity which expresses the principle that mass cannot be destroyed. Given the seven equations and the seven variables, most meteorological problems can be tackled on scales from dust devils to the global circulation. Of course, the relative importance of terms in each equation depends on the scale of the problem. So, for example, radiation in the heat equation is important on the largest scales, but not in the theory of the sea breeze or short-range weather prediction. On the other hand, vertical accelerations are important only for small-scale problems, e.g. cloud dynamics.

For many meteorological applications, some of the equations and variables are left out. For example, some short-range forecasting methods omit both moisture and the moisture equation. In some small-scale problems, air can be treated as incompressible so that density, temperature and moisture disappear as variables, and the thermodynamic equations can be omitted entirely. Since the equations of meteorology simplify the actual state of the atmosphere, we often refer to them as describing mathematical 'models' of the atmosphere, which describe atmospheric processes well, but not perfectly.

This chapter deals with some of the hydrodynamic equations. The remaining hydrodynamic equations and the thermodynamic equations are dealt with in Chapter 3.

The equation of continuity

The equation of continuity states that a given mass of air will remain the same mass, regardless of how much it moves about, or how much it is squeezed or stretched. In other words, it states the principle of conservation of mass. The way it is usually stated in meteorology is

Vertical convergence = horizontal divergence plus a density increase.   (2.1)

In practice, the density change is relatively small and can often be neglected. In that case, vertical convergence is essentially compensated by horizontal divergence. In mathematical terms:

$$\frac{\partial(\rho w)}{\partial z} = - \left[\frac{\partial(\rho u)}{\partial x} + \frac{\partial(\rho v)}{\partial y} \right], \qquad (2.2)$$

where z is the vertical co-ordinate, x and y are horizontal co-ordinates and u, v and w are velocity components. Figure 2.1 is an application of this principle. A tall, slender volume of air is subjected to vertical convergence. The result is horizontal divergence: the body becomes short and broad. The equation of continuity can be applied to any fluid, for example to a volume of custard. If you sit on the volume of custard, it will spread out sideways – it will diverge horizontally.

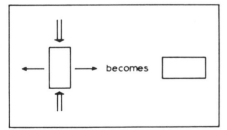

Fig. 2.1 Effect of vertical convergence.

We can apply the principle of continuity also to cyclones. In the region of a cyclone, air tends to flow towards the centre: we have horizontal convergence. Therefore, according to the equation of continuity, we must have vertical divergence. But the air cannot disappear into the ground; consequently, the air above the cyclone must rise (figure 2.2). Rising causes cooling and, frequently, condensation, so that we find precipitation associated with cyclones.

Since we can measure horizontal winds, but large-scale vertical motions cannot be measured directly (because they are very small), the equation of continuity can be used to infer the vertical motions from the horizontal wind field, as shown in figure 2.2.

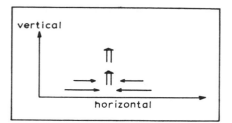

Fig. 2.2 Horizontal convergence in a cyclone (vertical velocities exaggerated).

The equations of motion

The equations of motion are derived from Newton's second law (see Atkinson, 1981). Newton's law can be stated as follows: if there are several forces, we can form the resultant of these forces (figure 2.3). This resultant force will produce an acceleration in the direction of the resultant. The magnitude of the acceleration will be proportional to the magnitude of the resultant force. For example, if there are two equal forces, one pulling air to the north and the other one to the west (figure 2.3), the air will accelerate towards the north-west. Of course, if all the forces cancel out, there will be no acceleration. But this does not mean that there will be no motion; it merely means that, when the forces are balanced, the body will move with a constant speed in a constant direction. For example, when a car moves at a constant 60 km h^{-1} along a straight highway, the propelling force and the retardation due to friction and wind resistance just cancel out.

If there are three forces, we can form the resultant of two of them. Then we form the resultant of this first resultant and the third force. This process can be extended to any number of forces.

When we talk about force in meteorology, we really mean force per unit mass of air. For example, when we say Coriolis force, we imply the Coriolis force on

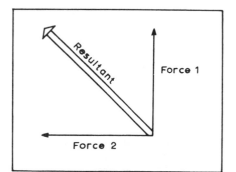

Fig. 2.3 Resultant of two forces.

one gram of air or on one pound of air, depending on what unit of mass we prefer. The advantage of forces per unit mass over ordinary forces is that Newton's second law becomes a little simpler; the law now states that the resultant of the forces per unit mass is equal to, not just proportional to, the acceleration.

Newton's second law can be applied separately in the vertical and horizontal, because a vertical force can never balance a horizontal force; neither can a horizontal force produce a vertical acceleration. Thus we can say, for example, that the sum of all the vertical forces equals the vertical acceleration.

We have already referred to the 'equations of motion' as the three components of Newton's second law, applied to the atmosphere or other fluids. One of the components is always taken in the vertical; different conventions are used for the selection of the two horizontal components. Frequently, we shall apply one component in the direction of the horizontal wind and the other at right angles to it.

Forces

Gravity

Gravity is actually the resultant of two forces, the gravitational attractive force of the earth for any object on or near the earth, and the centrifugal force due to the earth's rotation. The first of these forces is called gravitation. It is hundreds of times as strong as the centrifugal force, so that the forces of gravity and gravitation are not very different.

Gravitation, first discovered by Newton, is the universal force attracting any two bodies towards each other. The strength of this force increases as the mass of either or both of the bodies increases; and the strength decreases, in a 'squared' fashion, as the distance between the two bodies increases (for instance, doubling the distance decreases the force not to one-half, but to one-quarter).

At the Equator, the larger centrifugal force and the larger earth radius both act to make gravity weaker than at the poles; however, the overall difference is only about 0.5 per cent, which is often negligible in computations. The value commonly used, 9.8 m s^{-2}, is sufficiently accurate for most purposes.

Since gravitation decreases as the distance between two bodies — in this case the earth and the 'pancake' of air under consideration — increases, we see that gravitation and, consequently, gravity must decrease with height above the earth's surface. The value of gravity at 10 km, however, is only about 0.3 per cent different from the value at sea-level, and this difference is usually neglected in meteorological practice.

Pressure-gradient force

Wherever there are pressure gradients in the atmosphere, that is, variations of pressure with distance, air will experience a pressure-gradient force which tends to drive the air from the area of high pressure towards the area of low pressure.

The greater the variation of pressure with distance, the greater will be this force. Vertical pressure differences in the lower atmosphere are of the order of 100 mb km^{-1}, whereas horizontal pressure differences are of the order of 1 mb 100 km^{-1}. Consequently, the vertical pressure-gradient force is about 10 000 times the horizontal pressure-gradient force. This does not mean that the small horizontal pressure-gradient force can be neglected. In the horizontal, the other forces are also small. Even such small forces, acting for only an hour, can set up considerable wind. In the vertical, on the other hand, there is generally no strong net force since the vertical pressure-gradient force is balanced by gravity.

The reason for the large vertical pressure-gradient force is that gravity, which is itself large, compresses the atmosphere until the vertical pressure-gradient is sufficiently large to balance the compressing force, gravity.

The pressure-gradient force *PGF* from point A to point B may be given by

$$PGF = \frac{p_B - p_A}{D\rho}, \tag{2.3}$$

where p_B and p_A are respectively the pressures at the two points B and A located a distance D apart, and ρ is the mean air density between the two points. The force is always directed from higher pressure towards lower pressure. In exact mathematical notation:

$$PGF = -\frac{1}{\rho}\frac{\partial p}{\partial x}, \tag{2.4}$$

where x is distance.

The Coriolis force

Because our meteorological measurements are made on a rotating earth, the path of an object, which a stationary observer in space sees as a straight line, will appear curved to us on the rotating earth. If we see this path curve to the right, it appears as though a force has pulled the object to the right. The name given to the 'apparent' deflecting force is the Coriolis force.

We can illustrate this apparent force with a phonograph record rotating counterclockwise (figure 2.4). If a pencil follows a true straight line in space above the record from the centre C towards point B on the edge, while the record is rotating, the trace of the pencil will actually be curved on the record, somewhat like the curve CA in figure 2.4. It is as though a force had pulled the pencil towards A.

The deflection was not really caused by any force, but by the rotation of the record; nevertheless, motion in a rotating system when viewed *by an observer who is also rotating with the system* behaves as though there is such a Coriolis force. If the motion is counterclockwise, the deflection as seen in figure 2.4 is to the right; if clockwise, it is to the left. The Coriolis force (per unit mass) acting on a body has a magnitude equal to twice the rate of rotation of the system

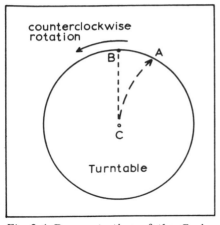

Fig. 2.4 Demonstration of the Coriolis force effect using a record turntable.

multiplied by the speed of the body:

$$\text{Coriolis force} = 2\omega V, \qquad (2.5)$$

where V is the speed of the body and ω is the rate of rotation of the system.

From the viewpoint of an observer standing at the South Pole, the earth is rotating clockwise once a day. Therefore, any object such as an air parcel leaving the South Pole will be deflected towards the left by a force per unit mass equal to $2\omega V$, where ω, the rate of rotation of the system, is now the rotation rate of the earth, one revolution per day. From the viewpoint of a person at the North Pole, the earth is rotating in a counterclockwise direction; hence a moving body will be deflected to the right by a force of the same magnitude.

Unfortunately for this discussion, we are also interested in atmospheric motions at locations other than the North or South Pole. To use the result that the Coriolis force is $2\omega V$ at these other latitudes, we must define ω more precisely as the rate of rotation of the horizontal plane about the local vertical direction (that is, pointing towards the centre of the earth), since we are interested in deflections of horizontal motions. We define the horizontal plane as the plane at right angles to the force of gravity (or the vertical) at the location of the observer; that is, the plane indicated by a carpenter's level.

At an arbitrary latitude, the horizontal plane spins about the vertical axis with rate of rotation $\omega \sin \phi$, where ϕ is the latitude. Thus, the general expression for Coriolis force is

$$CF = 2\omega V \sin \phi. \qquad (2.6)$$

This expression can be derived more rigorously by writing Newton's law for an observer outside the earth and then transforming it for use on the rotating earth (see Hess, 1959). In summary, the Coriolis force tends to pull a particle of air at right angles to its motion, to the left in the Southern Hemisphere and to the right in the Northern Hemisphere. The strength of the force depends on the speed of the air motion, the rate of rotation of the earth and on the latitude of the air parcel in question.

Friction

Consider a piece of paper on a table. Place your hand over this paper and pull the paper along. Why does the paper move? It moves because of friction between hand and paper. But what does this really mean? The hand is made up of a large number of molecules and, although these molecules have a great deal of erratic motion, there is an average motion of all the molecules in the direction that the hand is moving. The molecules of the paper before the force is applied have zero average motion. When the hand is applied to the paper, the molecules of the hand collide with the molecules of the paper, and some of the momentum (mass x velocity) of the hand is added to the momentum of the paper molecules*. Because of the gain of momentum, the paper begins to move. We call this phenomenon friction. In general, then, friction is a force between two objects brought about by molecular exchange of momentum between them.

Air molecules just above the ground move with a certain wind speed. Their random motions will exchange momentum with the ground, but, since the ground cannot move, the net effect of this exchange of momentum is to decrease the speed of the winds. Air molecules, however, are very inefficient in transferring momentum; in fact, it can be shown that, if molecules were the sole transferring agent, only the winds in the lowest 2 m would be affected noticeably.

Since the effect of 'friction' on winds is present to a much higher elevation, we must look for a different type of transferring agent, namely the vertical currents of the air itself. Momentum transfer by such vertical 'eddy' currents is many thousands of times more effective than momentum transfer by molecules in reducing the momentum of the air. If air is slowed down by eddy exchange of momentum rather than by molecular exchange, the force which is then retarding the air motion is called 'eddy' friction. Frequently, just the word friction is used, when eddy friction is implied. Friction is usually important in about the lowest 1000 m, a region which is therefore often termed the friction layer.

The equation of vertical motion

The hydrostatic equation

We noted above that the meteorological equations of motion relate the acceleration of an air pancake to the forces acting on the pancake.

In the vertical direction, both the Coriolis force and friction have extremely small components which can generally be neglected. Thus, there are only two

---

* Of course, some of the molecules in the paper try to slow down the hand, but the motion of the hand is too strong to be affected appreciably.

opposing forces in the vertical equation of motion: gravity and the vertical component of the pressure-gradient force. We can therefore state that, for most purposes:

Upward (vertical) acceleration = magnitude of upward vertical pressure gradient minus gravity.

Mathematically, $\dfrac{dw}{dt} = -\dfrac{1}{\rho}\dfrac{\partial p}{\partial z} - g,$

where w is the vertical velocity component and g is gravity. In many cases, particularly for large-scale synoptic applications, the vertical acceleration is quite small, and there exists a near balance between gravity and the pressure-gradient force. Such a balance is expressed by the 'hydrostatic' equation.

The equation is not balanced for small-scale motions. For example, in the theory of ocean waves or the theory of small wave clouds, the vertical accelerations of the individual particles are important. In words, the hydrostatic equation states:

Magnitude of gravity = magnitude of vertical pressure-gradient force.　(2.7)

In mathematical symbols

$$g = -\frac{1}{\rho}\frac{\partial p}{\partial z}. \tag{2.8}$$

Equations 2.7 and 2.8 show that the pressure falls more rapidly with increasing height in thick air (because it weighs more) than in thin air.

If we combine the hydrostatic equation 2.8 with the gas law, we find that pressure decreases more rapidly with height in cold than in warm air, again, because cold air is heavier. If we introduce a large amount of moisture (which is relatively light), the pressure decreases more slowly than in dry air at the same temperature.

Application of the hydrostatic equation

One of the uses of the hydrostatic equation is the determination of the height of a radiosonde; the radiosonde measures pressure, temperature and moisture. The only unknown remaining in the hydrostatic equation is height.

According to the hydrostatic equation, the decrease of pressure with height depends on the vertical distribution of temperature. In the 'standard' atmosphere, there is no moisture and the distribution of temperature is prescribed; e.g., its sea level temperature is $15°C$ and its lapse rate is $6.5°C\,km^{-1}$ in the troposphere. If this is fed into the hydrostatic equation, a 'standard' distribution of pressure with height results, which is used in pressure altimeters. These instruments sense pressure but read out height.

Hydrostatic stability

The hydrostatic equation states the condition for equilibrium between gravity and the vertical pressure-gradient force. But this can be stable or unstable. An equilibrium is stable when a system returns to its equilibrium once it has been disturbed. It is unstable if displacement is followed by further, more rapid displacement in the same direction.

It turns out from the adiabatic law (dealt with in more detail in the next chapter on thermodynamics) and the vertical equation of motion (including the acceleration terms) that displaced air will be accelerated away from its original location if the local lapse rate exceeds the adiabatic, $9.8°C \text{ km}^{-1}$. Such a situation is hydrostatically unstable and leads to strong vertical turbulence. In saturated air, the limiting lapse rate is the moist adiabatic. Conversely, if the temperature decreases slowly with height, or increases upward (inversion), the air is stable and vertical currents are damped.

This theory, however, omits the effect of wind. A strong wind shear (wind increasing or decreasing with height) can destabilize the air enough to make a hydrostatically stable layer actually unstable. This phenomenon is called 'dynamic' instability and is the cause of most clear-air turbulence, encountered particularly in the high troposphere.

The equations of horizontal motion

The horizontal components of Newton's second law are sometimes called wind equations. Two separate equations are written, one for the x-direction and another for the y-direction. Often, x is chosen west to east and y south to north, but we can sometimes simplify the discussion by measuring x at right angles to the flow, to its right, and y along the flow.

The only forces of importance in the horizontal are pressure-gradient force, PGF; Coriolis force, CF; and Friction, Fr. We will denote x-component and y-component by subscripts; acc stands for acceleration.

Then the wind equations are:

$$acc_x = PGF_x + CF_x + Fr_x, \tag{2.9}$$

$$acc_y = PGF_y + CF_y + Fr_y. \tag{2.10}$$

Of course, the acceleration terms measure the change of a wind component following the air. If we are interested in local wind changes, we will get additional terms due to advection of faster or slower air. These advection terms are 'nonlinear' (contain products of variables) and cause the major difficulty in the solution of the equations.

The geostrophic wind

Now, above the friction layer, friction can usually be neglected. Further, horizontal accelerations tend to be approximately 10 per cent or less of the Coriolis and pressure-gradient forces. Therefore, if we are satisfied to disregard the friction layer and areas of marked curvature in the trajectories and are willing to tolerate inaccuracies which, on the average, are of the order of less than 10 per cent, we may simplify the horizontal equation of motion by neglecting the acceleration and friction, thus leaving only the pressure-gradient and the Coriolis forces in balance with each other. We then deal with unaccelerated, frictionless flow, which is also known as geostrophic flow. Since any turning of the air in its motion would imply an acceleration, geostrophic flow is therefore, by definition, straight flow. Above the friction layer, the speed implied by this balance is usually a fairly good approximation of the true flow.

Serious exceptions occur when air flows rapidly around a curve; for example, in intense cyclones of all descriptions, and in curved flow in the upper troposphere where winds are fast. Thus, it should be noted that departures of the wind from geostrophic attend some of the most significant weather system developments. Additionally, at the Equator, where the horizontal Coriolis force is zero, the assumption of geostrophic flow cannot be used in its basic form.

If friction and accelerations are negligible, the pressure-gradient force must just balance the Coriolis force. This statement means not only that the magnitude of these two forces (vectors) is the same, but that their directions must be opposite. The pressure-gradient force is always towards low pressure at right angles to the isobars. The Coriolis force is always at right angles to the wind, and in the Northern Hemisphere is directed to the right. Therefore, since the two forces balance, the wind must be parallel to isobars, with low pressure to its left (figure 2.5). This statement is also known as the Buys–Ballot Law. In the Southern Hemisphere, the balance of Coriolis and pressure-gradient forces implies that the wind is still parallel to the isobars; but since the Coriolis force pulls to the left, low pressure is now to the right of the wind. The magnitude of the geostrophic wind is given from equation 2.9 if we line up the y-direction with the wind. In unaccelerated, frictionless flow, this equation then becomes:

$$0 = PGF_x + CF_x. \qquad (2.11)$$

Since here $CF_x = 2\omega V_g \sin \phi$ (V_g is the geostrophic wind speed) we can solve for the magnitude of V_g and find

$$V_g = \frac{PGF}{2\omega \sin \phi}. \qquad (2.12)$$

where V_g is the geostrophic wind. PGF_x has been set equal to the total pressure-gradient force because there is no pressure gradient in the y-direction. Thus, the

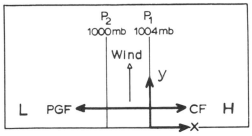

Fig. 2.5 Geostrophic wind balance in the Northern Hemisphere.

geostrophic wind is seen to be proportional to the magnitude of the pressure gradient.

Thermal wind

The geostrophic wind at a given level is proportional to the pressure gradient at that level. The wind at a higher level is proportional to the pressure gradient at this higher level. Therefore, the difference between geostrophic winds at two levels is proportional to the difference between pressure gradients at two levels. This wind difference is called the 'thermal' wind.

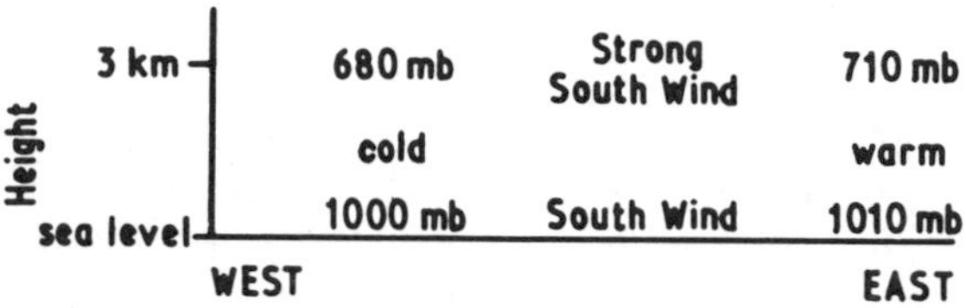

Fig. 2.6 Thermal wind.

In figure 2.6, we start at sea level, with pressure slowly increasing from west to east. This implies a geostrophic wind from the south (see also figure 2.5). Now, we suppose that there is cold air in the west and warmer air in the east. Then the pressure decreases most rapidly with height in the west, according to the hydrostatic equation. In our example, we have made the pressure 3 km above sea level 320 mb less than at sea level; in the warm air, the pressure difference is only 300 mb. As a result of the temperature difference, the pressure gradient at 3 km is stronger than at sea level, but in the same direction, so that the geostrophic wind at 3 km is also from the south, but much stronger than the surface wind. Thus, the horizontal temperature contrast has produced a vertical wind shear. The stronger the temperature gradient, the larger the wind shear; e.g., above fronts there are likely to be especially strong winds.

Quantitatively, the geostrophic and hydrostatic equations lead to the 'thermal wind' relations which can be summarized by the statements:

1. The vertical wind shear vector is parallel to the isotherms (on a horizontal surface). Warmer air is to the right, and colder air to the left of this wind shear vector (Northern Hemisphere).
2. The magnitude of the vertical wind shear is proportional to the magnitude of the horizontal temperature gradient.

Gradient wind

Considering figure 2.5, if we make the wind stronger than geostrophic, we increase *CF* (which depends on the wind) but do not change *PGF*. Therefore, there will be a net acceleration to the right. This follows also from equation 2.9. Hence, the flow will curve to the right (clockwise). In the Northern Hemisphere, this would imply anticyclonic flow. Hence, faster than geostrophic winds have clockwise trajectories. Or, in anticyclones, winds are faster than geostrophic. In cyclones, the winds are slower than geostrophic. For example, in a hurricane, the geostrophic wind may be 500 m s^{-1}, but the gradient (actual) wind only 75 m s^{-1} (150 knots)!

A mathematical analysis of gradient wind shows that anticyclonic wind speeds can never be very high, but cyclonic speeds are unbounded. This difference explains why anticyclones are never destructive.

Effect of friction on wind

In the lowest kilometre or so, wind is slowed down by friction. As it slows down, the Coriolis force decreases, and the pressure-gradient force turns the wind towards low pressure. This situation is shown in figure 2.7.

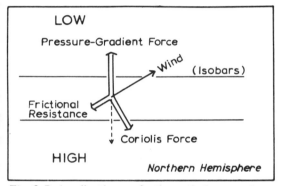

Fig. 2.7 Application of the wind equation near the ground.

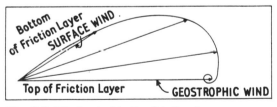

Fig. 2.8 Wind spiral in the friction layer.

Starting from the top of the friction layer, the winds get progressively slower and turn more towards low pressure as the ground is approached; the result is a wind spiral as shown in figure 2.8. The exact characteristics of such spirals depend on the conditions. For example, the thickness of the friction layer is much smaller at night than during the day; and the angle between surface wind and isobars is a function of many variables, particularly terrain roughness.

Friction is often introduced into complicated solutions of the meteorological

19

equations by simply assuming that the force is opposed to the wind and proportional to the square of its speed but the coefficient of proportionality is difficult to handle, depending again on terrain roughness, and also on the vertical temperature gradient near the ground.

References

Atkinson, B. W. (1981) 'Weather, meteorology, physics, mathematics', this volume, 1–7.

Hess, S. L. (1959) *Introduction to Theoretical Meteorology*, New York, Henry Holt & Co., Inc.

3
Atmospheric thermodynamics

H. A. PANOFSKY
Pennsylvania State University

The two previous chapters have shown the value of applying some of the principles of classical physics to the atmosphere. In this chapter we continue this application, considering relevant equations from thermodynamics. These include the equation of state (or gas law), a heat equation and an equation expressing conservation of moisture.

The equation of state (Gas law)

The equation of state for dry air (that is, air containing no water substance) is

$$\rho = p/RT. \tag{3.1}$$

Here, R is a constant, ρ the air density, T the temperature and p the pressure. Since T and p are usually observed, the equation is generally used to determine or to eliminate the density. Qualitatively, increasing the temperature expands the air and decreases the density; increasing the pressure compresses the air and increases the density. The equation of state for dry air is quite accurate; however, when particularly precise calculations are to be made, a correction has to be made for moisture. The molecular weight of water vapour is only 18, as compared to 29 for dry air. Therefore, replacing some of the dry air with water vapour will decrease the density. Roughly, 1 per cent of moisture decreases the density by 0.6 per cent, and 0.5 per cent of moisture, by 0.3 per cent. These are rather typical concentrations of water vapour, and it is obvious that they produce only small changes in density.

We often state that the equations such as the gas law are applied to 'pancakes' of air, since the volumes we consider are usually thin in the vertical but large horizontally.

The first law of thermodynamics

The first law of thermodynamics states:

$$\text{Heat added} = \text{increase of internal energy} \tag{3.2}$$
$$+ \text{work done against outside pressure.}$$

In mathematical terms, this equation is often written:

$$dh = de + pd\alpha, \qquad (3.3)$$

where dh is a small amount of heat added to unit mass (e.g. a gram) of air, de the change in internal energy, p the pressure and $d\alpha$ the change in volume of this mass of air. This means that if heat, which is a form of energy, is added to a pancake of air through its boundaries, this energy is used in two ways: either the internal energy of the air can be increased, or the air can expand against the surrounding outside pressure, or both.

Energy can be added from the outside by processes such as radiation, or mixing with warmer air along the boundaries. A change of internal energy means that either the potential or the kinetic energy of the molecules changes. Air is nearly a perfect gas, which behaves as though all the molecules are like billiard balls. They move as though there are no forces between them except collisions which preserve energy and momentum. Relative to each other, individual molecules have no potential energy. Hence the only important form of energy of the molecules of a perfect gas is kinetic energy – the energy due to the motion of the molecules. The kinetic energy of the molecules of a gas is indicated by its temperature; therefore, the only way in which a perfect gas can increase its internal energy is by increasing its temperature. That is why we find de is proportional to the temperature change dT.

The only important way in which a gas can do work against outside pressure is by expansion. Therefore, we can restate the first law of thermodynamics as follows: if we add energy to a mass of air through its boundaries, this energy is used to raise the temperature of this body of air, or to expand this body of air, or a combination of these two alternatives.

The dry adiabatic process

Many of the processes which add heat energy across the boundary of a pancake of air are quite slow. Therefore, it has been useful in many problems to consider a process in which no energy is added, and in which evaporation and condensation are unimportant. This process is called the dry adiabatic process. The word 'adiabatic' is derived from three Greek roots: a, the negative (as in atheist); dia, through (as in diameter); and batic, walking. Thus, 'adiabatic' means 'not walking through'. Of course, the energy is the quantity that does not walk through.

If there is no energy flow across the boundary of the pancake, the first law of thermodynamics states:

> Energy for temperature increases plus energy   (3.4)
> for expansion equals zero.

Transposed, this means that energy for expansion equals negative energy for temperature increase, or:

> Energy for expansion equals energy released   (3.5)
> by temperature decrease.

In mathematical terms, equation 3.5 is written

$$pd\alpha = -c_v dT, \qquad (3.6)$$

where c_v is the specific heat of a constant volume. In particular, if air rises, energy is required for its expansion. But where can this energy come from? No energy is available from outside or from condensation. Therefore, the energy must come from kinetic energy (energy of motion) of the molecules making up the gas. The molecules slow down. But, since the temperature is a measure of the kinetic energy of the gas, the temperature must decrease. Thus, expansion of air (or lifting it) cools the air. This effect can be seen, for example, by punching a hole into a bicycle tyre: the air expands and is quite cool. Conversely, sinking air is compressed, and compression produces warming, a fact that is easily noticed by the increasing temperature of a bicycle pump in action.

Quantitatively, the dry adiabatic temperature change can be computed from:

$$\text{Increase of temperature} = \frac{\text{Increase of pressure}}{10\ 000\ \text{times air density}}, \qquad (3.7)$$

provided the increases (or decreases) in temperature and pressure are small, relative to the initial values.

In equation 3.7 the pressure is in millibars, the temperature is in degrees Celsius and the density is in grams per cubic centimetre. This density unit is chosen so that it equals one for water. Since the density of air near the ground is about 1/1000 that of water, equation 3.7 states that: near the ground, a pressure decrease of 10 mb requires a $1°C$ temperature decrease if the process is dry adiabatic. The pressure near the earth's surface decreases in the vertical approximately 10 mb in 100 m, so that, if air near the ground is lifted adiabatically, it will cool $1°C$ for each 100 m lifted. Conversely, adiabatic sinking by 100 m is accompanied by a $1°C$ warming.

At large heights (near 20 km), the density is about 1/10 that at the surface. At these altitudes, it will only require a lifting through a single millibar to produce a cooling of $1°C$; but the isobaric surfaces are much further apart at such heights, as will be seen later. Thus, it still takes a lift of approximately 100 m to produce $1°C$ cooling at all heights.

Poisson's equation

We can use the gas law ($p = R\rho T$) to eliminate the density from equation 3.7. This, plus the mathematical operation of integration, leads to Poisson's equation:

$$(T_2/T_1) = (p_2/p_1)^{0.287}. \qquad (3.8)$$

Here, p_1 and T_1 are the initial pressure and temperature, and p_2 and T_2 are the final pressure and temperature. The units of p are arbitrary, but T must be in degrees Kelvin, defined as $°C + 273°$. The exponent 0.287 is nearly equal to 2/7, a value which is often used. In practice it is generally inconvenient to substitute

Dynamical Meteorology

numbers into equation 3.8 and the use of diagrams is preferred for the application of Poisson's equation.

The adiabatic approximation is used in a great deal of theoretical work, particularly in numerical weather prediction. It is strictly correct only for dry air; but is also an excellent approximation in humid air, as long as the air is not saturated. The advantage of equation 3.8 over equation 3.7 is that it can be used for large pressure and temperature differences. For example, suppose a pancake of air has a pressure of 1000 mb and a temperature of 20°C (293°K). What will be its temperature if this pancake is lifted to 700 mb? The answer can be found by solving equation 3.8 for T_2: 265°K, or −8°C.

Restrictions of the adiabatic approximation

Since the adiabatic approximation is so often used (the word 'dry' is commonly omitted), it is important to realize its restrictions. To understand what these restrictions are, we should first realise how fast air temperature is likely to change due to large-scale adiabatic processes. Large-scale vertical motions are of the order of a centimetre per second. Therefore, if air rose or descended throughout a day, adiabatic cooling or warming would be of the order of 9°C during this period. This is much larger than temperature changes due to absorption or emission of radiation. Radiation produces changes in the free air of about 1°C day^{-1}, although the change is much larger just above the ground. Clearly, radiation is less important than adiabatic temperature changes except close to the ground, provided that we have such vertical motions. However, we cannot maintain such large organized vertical motions over prolonged periods. Air that goes up must eventually come down. Thus, over a long period of time, the average temperature change (due to vertical motion) may be smaller than that due to radiation. In fact, it is basically the radiation from the sun that drives the atmosphere. The adiabatic approximation is valid only for atmospheric changes over a few days, at most.

Air is a very poor heat conductor. This explains why a double window keeps out the cold so much more effectively than a single window. Nevertheless, the heating of air by heat conduction cannot be neglected in the lowest few millimetres. Above that, molecular heat conduction is of little importance.

There is still another diabatic process which can be of great importance. (Diabatic is the opposite of adiabatic.) This other process is mixing. One way a large, hot pancake of air can cool is by being surrounded by cooler air; small-scale motions — sometimes called eddies — will mix this colder air with the warm air in the pancake. This process can occur everywhere, but is particularly important in the lowest kilometre of the atmosphere where large-scale organized vertical motions are suppressed (except near mountains) — and adiabatic warming and cooling must be relatively unimportant. Still, every clear afternoon the air becomes warm, especially in the lowest 300 metres. The cause is mixing. First, the radiation from the sun is absorbed by the ground. Next, the ground heats up, warming the layer of air immediately above it by molecular conduction. The heating of the air sets off convection currents, with horizontal dimensions of perhaps a kilometre and slightly smaller vertical dimensions. These convective eddies carry the heat from the ground to higher levels. Thus, the air is being heated diabatically. The cooling at night occurs chiefly because the water vapour in the atmosphere loses heat by infrared radiation. Again, this is a diabatic process.

Finally, the adiabatic equations are not accurate when condensation or evaporation is important; for a little evaporation or condensation involves a great deal of energy. The consequences will be discussed later. In summary, the adiabatic equations are good approximations unless we are considering air near the ground, long-period developments or processes in which condensation or evaporation is important.

Potential temperature

Let us suppose that we have some dry air with pressure p and temperature T. We wish to calculate the temperature which this air will have when it is brought adiabatically to a pressure of 1000 mb. This temperature is defined as the 'potential' temperature and given the symbol θ.

We can find θ by setting p_2 equal to 1000 mb in equation 3.8 and then solving for T_2 which is then equal to θ. Thus we obtain:

$$\theta = T \left(\frac{1000}{p}\right)^{2/7}. \tag{3.9}$$

Therefore θ, the potential temperature, is the temperature which dry air will have if brought adiabatically to 1000 mb. It is a function of T and p only, as seen by equation 3.9, and is usually given in degrees Kelvin.

In order to understand better the usefulness of this potential temperature, we now construct a graph (figure 3.1). Here, temperature is plotted on the abscissa

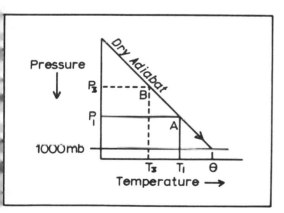

Fig. 3.1 Schematic graph of potential temperature.

Dynamical Meteorology

(horizontal axis) and pressure is the ordinate (vertical axis). In meteorological diagrams the pressure scale usually increases downward, because in the actual atmosphere the highest pressure is at the lowest level. We consider a point A which has pressure p_1 and temperature T_1. We now draw a line connecting all points which can be reached from point A through an adiabatic process. This is known as the dry adiabat through A. According to the definition of θ, this adiabat intersects the 1000 mb line at a temperature equal to the potential temperature.

Now suppose that the pancake of air with pressure and temperature of point A is lifted adiabatically, arriving at a higher level denoted by point B with pressure p_3 and temperature T_3. A and B are on the same adiabat and have the same potential temperature. Thus, the potential temperature of the pancake remains the same when the pancake is raised (or lowered) adiabatically. We say that the potential temperature θ is a conservative property for adiabatic processes. This means we can use θ as an identifier. As long as there is no significant mixing, radiation, evaporation or condensation, the pancake will retain its θ − regardless of how much it ascends or descends. Of course, mixing with warmer air will increase θ, and the emission of radiative energy can lower θ. In general, the potential temperature, θ, of a pancake of air increases when heat is added to the pancake, and it decreases when it loses heat.

In the first part of this treatment of atmospheric thermodynamics, we dealt with dry air. We now turn to some aspects of the behaviour of moist air.

Properties of water vapour

Even though water vapour never amounts to more than 4 per cent of the air − and is usually very much less − it is of great importance for two reasons:

1. A little vapour can absorb or emit a great deal of radiation in the infra-red wavelengths.
2. Water vapour can change to liquid or to ice, and the amount of heat energy involved in these phase changes is substantial, even when the amounts of vapour involved are small.

In order to be more precise about the properties of water vapour, we will first introduce the vapour pressure, e. This is merely that portion of the total air pressure which is due to water vapour, and is usually measured in millibars. For example, if the total atmospheric pressure is 1000 mb, the vapour pressure may be 10 mb − the vapour is responsible for merely 1 per cent of the total pressure.

Now consider a water surface at a temperature of $10°C$. The air above this surface contains some water vapour. Let us suppose that the vapour pressure of this vapour is e. At all times, there is an exchange of water molecules between the liquid and the air (figure 3.2); water drops are evaporating and joining the air, and water vapour molecules are condensing and joining the liquid. If e is very small, more water will evaporate than will condense; and e will be increasing (figure 3.2a). If e is very large, there is more condensation than evaporation; and *e* will decrease (figure 3.2b). In both cases, a state of equilibrium is eventually reached when the number of evaporating and condensing molecules is equal. In this state we find that the vapour pressure is the equilibrium vapour pressure (figure 3.2c). We will denote the equilibrium vapour pressure by e_s. If *e* is greater than e_s, there is net condensation; if *e* is less than e_s, there is evaporation. e_s is also called the saturation vapour pressure; the reason for this will appear later.

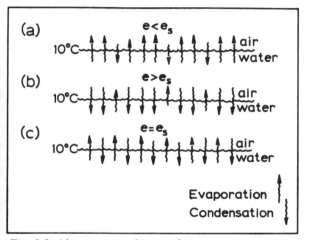

Fig. 3.2 Air—water exchange of water vapour.

It can be shown from the first law of thermodynamics that, over a flat surface of pure water, e_s depends only on temperature, increasing as the temperature increases. This relationship is quite complicated; therefore, it is given here in tabular form (table 3.1).

Condensation and precipitation

Let us assume that *e* is gradually increased by adding water vapour to a pancake of air. When will condensation begin? We must ask first on what the vapour will condense. The atmosphere is full of impurities – sea salt, sand, dust, industrial products, etc. These microscopic particles are called condensation nuclei, since condensation of water vapour will occur upon them if they are present in the right size and quantity.

The saturation vapour pressure values in table 3.1 are strictly correct for flat surfaces and pure water. e_s is larger over curved droplets than the values given in table 3.1. e_s is also smaller over impure droplets; these impurities provide condensation nuclei. Thus, condensation occurs on condensation nuclei before *e* reaches e_s. However, *e* will eventually have to equal e_s in order for the drop to grow large enough to produce clouds and precipitation; for the drop is then so large that impurities and curvature are unimportant. Therefore, for practical purposes, we can say that condensation occurs when e reaches or just surpasses e_s. The vapour becomes saturated at that time, and large drops can form. For this reason, e_s is usually called the saturation vapour pressure.

At temperatures below freezing, water may exist in the solid form (ice) or in liquid droplets (supercooled water). As shown in table 3.1, the saturation vapour

Table 3.1 *Saturation vapour pressure (mb) at various temperatures*

Over water						Over ice	
$T(^\circ C)$	e_s(mb)	T	e_s	T	e_s	T	e_s
100	1013.25	50	123.4	0	6.1	0	6.1
95	845.3	45	95.9	−5	4.2	−5	4.0
90	701.1	40	73.8	−10	2.9	−10	2.6
85	578.1	35	56.2	−15	1.9	−15	1.6
80	473.7	30	42.4	−20	1.3	−20	1.0
75	385.6	25	31.7	−25	0.81	−25	0.63
70	311.7	20	23.4	−30	0.51	−30	0.38
65	250.2	15	17.0	−35	0.31	−35	0.22
60	199.3	10	12.3	−40	0.19	−40	0.13
55	157.5	5	8.7	−45	0.11	−45	0.07
				−50	0.06	−50	0.04

pressure e_s is larger over supercooled water than over ice. This difference has an important consequence: when water drops and ice crystals exist together in a cloud, the vapour between these particles cannot be in equilibrium with both at the same time. Let us suppose that the water vapour is initially in equilibrium with the ice crystals. Then e is too small for equilibrium with the water, which will therefore begin to evaporate. As a result, e becomes too large for equilibrium with the ice, and condensation will occur on the ice crystals. Thus, the crystals will grow at the expense of the water drops. If sufficient moisture and growth time is available, the crystals will grow large enough to form precipitation. This mechanism forms the basis of the Bergeron–Findeisen theory of precipitation formation within clouds.

Relative humidity and dew point

The quantity $100\, e/e_s$ (in per cent) is called the relative humidity. It describes how close the air is to saturation. For example, if e is 5 mb and e_s is 10 mb, the relative humidity is 50 per cent, i.e., the air is 'half saturated'. There are basically two different ways in which the air can be brought to saturation: by adding more vapour to the air, which increases e, or by cooling the air, which decreases e_s. We sometimes say – not very precisely – that cold air cannot hold water vapour as well as warm air.

Saturation by cooling is a more important process than saturation by adding vapour. For example, almost all clouds are produced by the lifting of moist air, which cools adiabatically and condenses. Early morning fog is also due to cooling of the night air until e_s becomes as small as e.

A measure of moisture closely connected to e and the relative humidity (r.h.) is the dew point T_d. The dew point is defined as the temperature at which water vapour in the air will reach saturation if the air is cooled without a change of pressure, provided that no vapour is removed from or added to the air. The dew point is essentially the temperature at which dew forms when air cools during the night.

In order to see how the dew point can be computed, consider the following example: suppose that the vapour pressure in the air is 12.3 mb and the temperature is 15°C. Table 3.1 shows that e_s at this temperature is 17.0 mb. Hence, the air is not saturated. Since the vapour pressure is only 12.3 mb, the vapour will condense after it has been cooled to that temperature for which 12.3 mb indicates saturation. Table 3.1 shows that this is 10°C. Therefore, the dew point is 10°C. Note that we did not use the temperature value of 15°C to determine the dew point. All that is necessary is to look up the temperature from table 3.1 at which 12.3 mb indicates condensation. Generalizing from this computation, we may state that the dew point T_d has the same relation to the vapour pressure e as the temperature has to the saturation vapour pressure e_s. Table 3.1 gives both relationships.

Specific humidity and mixing ratio

Vapour pressure, relative humidity and dew point depend only on the conditions of the water vapour. The values for these quantities are independent of the amount of dry air present. There are other important moisture variables which involve the air. The most important of these, for theoretical purposes, are the specific humidity (q) and the mixing ratio (w).

The specific humidity is the ratio of the density of the water vapour in a given volume to the density of the total air (including water vapour) in that volume. It is also the ratio of the mass of water vapour in a given volume to the mass of air in the given volume. The mixing ratio is the ratio of the density (or mass) of water vapour to the density (or mass) of dry air only (i.e., excluding water vapour). Since the density of dry air is very nearly equal to the density of the total air, the difference between specific humidity and mixing ratio is often neglected.

The mixing ratio and specific humidity depend not only on the vapour pressure, but also on the total pressure. Approximately, both are given by:

$$q = 0.62 \, e/p, \qquad\qquad (3.10)$$

an equation derived from the gas laws, applied to dry air and water vapour. Given the dew point, one may use this equation to calculate w or q. In that case, e is found from table 3.1 and q from equation 3.10.

The mixing ratio (also the specific humidity) is important because it is often conservative: as air moves, the mixing ratio remains constant. Only mixing with drier or wetter air, evaporation or condensation will change the mixing ratio.

As we have seen, when e reaches e_s the air is saturated, and condensation is to be expected. At this point, the specific humidity, q, reaches the value

$$q_s = 0.62\, e_s/p, \qquad (3.11)$$

which is called the saturation mixing ratio. For practical purposes, q does not exceed q_s for when q reaches q_s, condensation begins to remove some of the water vapour from the air.

Since e_s depends only on temperature (table 3.1), q_s depends only on pressure and temperature (equation 3.11). q_s is a few per cent at 1000 mb in very hot air, but less than 1/10 of 1 per cent in very cold air. Again, we see that 'cold air cannot hold much water vapour'.

If we divide equation 3.10 by equation 3.11, we obtain an expression for the relative humidity:

$$\text{r.h.} = 100(q/q_s) \text{ per cent.} \qquad (3.12)$$

Equation 3.12 again shows that saturation can be reached either by adding moisture (increasing q) or by cooling (decreasing q_s). Equation 3.12 is particularly useful for operations on thermodynamic diagrams, such as the tephigram.

Phase changes

Next, we will comment on changes of phase of water. The process of changing liquid to vapour is called evaporation; the process of changing vapour to liquid is called condensation. Approximately 600 calories are required to evaporate 1 gram of water at the usual meteorological temperatures. The effect can be felt when one comes from a swimming pool and is exposed to the air; evaporation of water from the body produces cooling, since it removes energy from the motion of the molecules in the body. Most evaporation takes place at the surface of the earth – from seas, lakes, vegetation or moist ground. This may saturate the layer of air immediately above the surface; the vapour is then introduced into the air above by mixing.

Condensation adds heat to the atmosphere. It occurs most frequently whenever clouds are formed due to cooling of rising air, so that q_s decreases to q. This process generally takes place well above the earth's surface. Of course, condensation can also occur near the ground, as in the case of fog.

It is also possible to transform ice into vapour and vapour into ice without going through an intermediate liquid stage. Cirrus clouds are made of ice crystals which grow directly from water vapour by crystallization. In this process, the heat of sublimation is released – about 680 calories for every gram of air. Conversely, ice or snow can change directly into vapour. A layer of snow on the ground can disappear without melting; the snow is transformed into vapour. The heat required for this process (sublimation) is again the heat of sublimation, approximately 680 calories per gram.

The moist adiabatic process

We have noted that the dry adiabatic equation is approximately correct even for humid air, as long as it is not saturated. If the air is saturated and condensation occurs, we must allow for the heat of vaporization*. When saturated air is lifted, the energy needed for the expansion is partially supplied by the heat released as the vapour condenses. Since the heat of vaporization is never quite enough to accomplish the expansion, heat must still come from the kinetic energy of the air molecules themselves. Thus, saturated air will cool when lifted, but much more slowly than in the dry adiabatic process. The process in which rising air is cooled in spite of the addition of heat of vaporization is called the moist adiabatic process.

The effect of condensation is particularly important when condensation occurs at high temperatures; for then the amount of vapour condensed is relatively great, and the heat added to the air is substantial. Quantitative expressions for the moist adiabatic process are quite complex; most problems involving this process are handled with thermodynamic diagrams.

Among the several moist adiabatic processes that are defined, the following two are most often used: the reversible saturation-adiabatic processes, in which the condensed vapour is retained in the air; and the pseudo-adiabatic process, in which all the condensation products fall out. Because the mass of liquid involved (which makes the difference between the two processes) is so small compared to the mass of air, the difference in cooling between these two processes is usually neglected.

An additional complication arises in the mathematics of the moist adiabatic process from the fact that the amount of heat released during the lifting depends on whether the condensation product is ice or liquid; more heat is released when ice is produced. Whether liquid or ice is produced in the rising air depends on factors besides just temperature. Clouds are often composed of liquid water, even at temperatures well below freezing. The temperature at which freezing occurs in a cloud depends on the droplet size, on the amount of impurities present and, perhaps, on other factors. Therefore, it is impossible to use the latent heat of vaporization over one specific temperature range and the heat of sublimation for another range. Sometimes it is simply assumed that the condensation products remain liquid indefinitely. When the air is so cold that the clouds are chiefly in ice form, the amount of water is so small that the difference in heat released in the two processes is unimportant.

The moisture equation

The moisture equation states that moisture cannot be gained or lost, but may change phase. In other words, then, it states that change of moisture with time equals gain by evaporation minus loss by condensation, plus loss or gain by mixing. The gain occurs mostly near the ground by evaporation into the lowest layer, followed by mixing. Condensation generally occurs in the free atmosphere, whenever the relative humidity approaches 100 per cent. In mathematical terms,

* Some authors call this the heat of 'condensation'.

equation 3.3 reads:

$$\frac{dq}{dt} = \left(\frac{dq}{dt}\right)_{evap} + \left(\frac{dq}{dt}\right)_{conden} - \frac{1}{\rho}\frac{\partial F}{\partial z}, \qquad (3.13)$$

where F is the *vertical* transport of moisture due to mixing and z is height. Horizontal mixing is neglected in large-scale problems. However, in problems on smaller scales, such as cloud dynamics, terms analogous to the last term in equation 3.13 in *horizontal* directions become important.

The concept of advection

Most of the meteorological equations deal with changes of some property, such as temperature, moisture, or wind, for a particular 'parcel' or 'pancake'. In mathematical terms, this fact is indicated by use of the 'total' or 'substantial' derivative, dX/dt, which gives the change with time of any property X, of a parcel, following that parcel. For applications of the equations, either on a computer, or for other purposes, it is awkward to follow parcels. It is much easier to work with local change, that is, change at a fixed point. A local change of X with time is indicated by the partial derivative, $\partial X/\partial t$.

When local changes are to be estimated, additional terms appear in some of the equations. These additional terms express changes 'due to advection' which express changes due to new air replacing the old air at a given place, because of horizontal or vertical air motions.

For example, if hot air is located to the south, a south wind leads to warming. If dry air is located above, sinking will decrease the specific humidity. There would be no advective changes if there was no motion, or if the meteorological properties were the same everywhere. Therefore, advective terms of property X depend on gradients of this property, and on velocity components. In fact, in all such terms, gradients are multiplied by velocity components. In particular, a change of temperature due to south winds is indicated by the term $-v\partial T/\partial y$, appearing in the heat equation. Here, v is the south–north wind component, and $\partial T/\partial y$ the south–north temperature gradient. For example, if v is 10 m s^{-1} and the temperature gradient is $-1°C\,(100$ km$)^{-1}$, the temperature change at a point due to advection is:

$$\frac{\partial T}{\partial t} = -(10 \text{ m s}^{-1})\left(-\frac{1°C}{100\,000 \text{ m}}\right) = 1°C\,(10\,000 \text{ s})^{-1}$$

or a warming of $1°C$ in about 3 hours. (Note that $\partial T/\partial y$ is negative because temperature *decreases* with *increasing y*, the south–north co-ordinate.)

4

Atmospheric vorticity and divergence

R. S. HARWOOD
University of Edinburgh

Vorticity is a measure of the local spin of a portion of fluid and divergence is a measure of whether the portion is expanding or contracting. Those will hardly do as definitions, so we will be more precise in a moment, but it should suggest that these two properties are potentially useful and at the same time fairly straightforward (even if the names may be off-putting and they are usually confined to texts crammed with complicated mathematical symbols). Their great utility arises from (1) the interaction between them and (2) the close relationship between the divergence and the vertical velocity. In particular, the local spin of fluid particles relative to the earth has opposite signs in cyclones and anticyclones and divergence is the principal agency responsible for the vorticity change which a particle must experience as it flows from one to the other. We shall show below that, by observing vorticity changes, we may deduce the vertical velocity, using divergence as the intermediate step. This is a valuable inference, for the vertical velocity, whilst crucial to the formation or suppression of cloud and rain, is difficult to measure directly, being very small. Unfortunately, if we wish to do anything quantitative or wish to be certain that our ideas are rigorous, we usually have to express our ideas in mathematics, which may appear daunting, but actually makes life easier. 'The use of mathematics in science is that of a language in which we can state relationships too complicated to be described except at inordinate length in ordinary language' (Jeffreys and Jeffreys, 1946). As this chapter is intended as an introduction to the ideas of vorticity and divergence for non-mathematicians, some statements will have to remain unproved and be taken on trust if it is not to assume inordinate length, but the statements should at least be plausible. First, we shall say what is meant by divergence, exploring its relationship to vertical velocity. Second will come a discussion of what vorticity is and how it might be measured. Third, we shall consider how divergence changes vorticity. Finally, we shall apply the ideas to important synoptic-scale weather systems.

Divergence

We shall begin by discussing divergence. As we wish to obtain a measure of whether fluid elements are expanding or contracting, it is natural to imagine marking a piece of fluid (perhaps with dye if it is a liquid, or smoke if it is a gas) and watching what happens to it. Let us confine ourselves to air and we will imagine that we can instantaneously release a horizontal, circular pattern of neutrally buoyant smoke. As it blows downstream with the wind it will deform as sketched in figure 4.1 (possibly ceasing to be horizontal). If we use A to denote the area of the smoke curve projected on to a horizontal plane expressed as a fraction of the original area, then A will vary with time as shown in figure 4.2, having a value of one at the moment of release, but subsequently becoming either greater than one, in which case the motion is said to be horizontally *divergent*, or less than one, in which case the motion is said to be horizontally *convergent*. Figure 4.1 has been drawn to represent divergent motion.

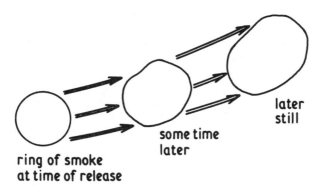

**ring of smoke
at time of release**

**some time
later**

**later
still**

Fig. 4.1 Plan view of the progress of a smoke curve which is circular at the time of release. As the horizontal area enclosed by the smoke is increasing, the motion is divergent.

To assign a value to the divergence we shall find the instantaneous rate of increase of A at the time of release. Inspecting figure 4.2 makes it clear that we need the slope of the tangent at the release time of the curve of A against time. In figure 4.2 this slope has been drawn to give an increase of A of 40 per cent in two days, i.e. a slope of 0.2 (day^{-1}) or $2 \times 10^{-1} \times (24 \times 3600 \text{ s}^{-1}) = 2.3 \times 10^{-6}$ s^{-1}. We could call this slope *the divergence* were it not for the slight complication that, since the flow properties vary in space, the answer depends to some extent on the size of circle we started with, even though we are concerned only with fractional rates of increase. It is also obvious that we should try to define the divergence at a point, i.e. the circle should be made as small as possible. Fortunately, provided the flow is smooth we can actually construct a value of the slope for a circle of zero size. All we have to do is to choose a point, find the instantaneous fractional rate of increase for several circles of different diameters centred

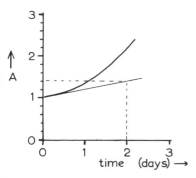

Fig. 4.2 In this figure, A is the ratio of the area of the smoke curve of Fig. 4.1 to its area at the time of release. This changes in time as shown by the thick curve. The thin straight line shows the initial rate of change of A which is $(1.4 - 1)/(2 \text{ days}) = 2.3 \times 10^{-6} \text{ s}^{-1}$. If the initial area was sufficiently small this number is called the divergence.

on it and then extrapolate to the value for a circle of diameter zero. That value is what we shall call the divergence. If the value turns out to be negative we may instead call it convergence. For example a *divergence* of $-1.6 \times 10^{-6} \text{ s}^{-1}$ would be called a *convergence* of $+1.6 \times 10^{-6} \text{ s}^{-1}$ and vice versa. In practice, we shall not need to extrapolate to zero radius; it will suffice to find the fractional rate of increase of a finite circle provided it is not too large to begin with.

We could have devised a measure of three-dimensional expansion or contraction based on rates of increase in volume as is the usual practice in fluid dynamics. Strictly, therefore, we ought always to specify *horizontal* convergence or divergence for our measure. However, the two-dimensional version is by far the most useful in synoptic meteorology and the word horizontal is understood.

Measuring the divergence

A convenient way of making a practical measure of divergence is to use a regular array of grid points like that in figure 4.3a which are assumed to be on a map with almost any type of projection so long as it is conformal. Let us use the symbol u_G for the component at G of velocity in the direction GC, which we shall call the x direction, and u_C for the component at C in the same direction. Let v (with the appropriate subscript) be the component of velocity in the direction EA. This is explained in more detail in figure 4.3b. Then providing d is small enough,

$$\text{divergence at O} = \frac{u_C - u_G}{2d} + \frac{v_A - v_E}{2d}. \tag{4.1}$$

The usual symbol for the divergence of the horizontal wind on the left-hand side of equation 4.1 is div $\mathbf{v}_h$. Those who are familiar with partial derivatives will recognize that as d gets smaller the right-hand side approaches $(\partial u/\partial x) + (\partial v/\partial y)$ so really equation 4.1 is an approximation to

$$\text{div } \mathbf{v}_h = \frac{\partial u}{\partial x} + \frac{\partial v}{\partial y},$$

which is the mathematician's definition of divergence.

The reader will be able to see for himself that the right-hand side of equation 4.1 measures just the kind of things we discussed in the definition except that we are using a square not a circle. Thus $u_C - u_G$ is the rate at which the square B DFH is extending in the x direction and $2d$ is its original length in that direction.

There are a number of remarks which should be made about this formula. The first is probably self-evident but we shall state it nonetheless. Namely, that we are talking about velocity components not just the magnitude of the winds so that if, for example, GC happened to lie west–east and the wind at B was 20 m s^{-1} *from* $30°$ north of west, v_B would be $-20(\sin 30°) = -10 \text{ m s}^{-1}$ (see figure 4.3b) and it is important to have the minus sign. The second remark is that, although in principle, we should have d as small as possible, in practice it will need to be several tens of kilometres if the errors in estimating the velocity components are to be small compared with their differences. The third remark is that the divergence is extremely tricky to estimate, even using this formula, because usually the value of $(u_C - u_G)/(2d)$ is very nearly equal to $(v_A - v_E)/(2d)$ but has the opposite sign, so you have to know both of them very accurately to get a reasonable estimate of the total. The fourth remark is closely related to the third. It is that the wind components must be calculated from the real winds

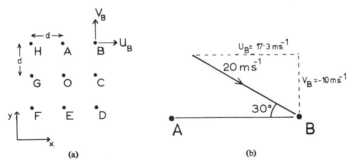

(a) (b)

Fig. 4.3(a) A grid of points used for working out divergences and vorticities. d is the distance apart of the points on the Earth. u is the component of wind in the x-direction and v the component in the y-direction. The subscript shows at which point the component is measured.

Fig. 4.3(b) The meaning of the wind components. The observed wind in this example is 20 m s^{-1} making an angle of 30 degrees with the direction AB. So $u_B = 20 \cos 30° = 17.3 \text{ m s}^{-1}$, and $v_B = -20 \sin 30° = -10 \text{ m s}^{-1}$. The minus sign appears because v_B is positive in the direction CB, whereas this component is in the direction BC.

and not from the geostrophic winds because, for geostrophic winds, the cancellation between $(u_C - u_G)/(2d)$ and $(v_A - v_E)/(2d)$ is exact (except for a small effect due to the change in Coriolis parameter with latitude): expressed differently; *the main contribution to the divergence comes from the ageostrophic part of the wind.* The reader may like to prove for himself the cancellation for geostrophic winds. All that is necessary is to note that the geostrophic estimate for u_C is $(p_D - p_B)/(\rho f\, 2d)$ where p_D is the pressure at D, f is the Coriolis parameter, and ρ the density. Treating ρ and f as constants he will have no difficulty writing down similar expressions for u_G, v_A and v_E, and obtaining the result.

There is another formula for computing the divergence based on the rate of change of velocity along a streamline and the rate of change of the separation of streamlines. However, since the difficulty of obtaining sufficiently accurate winds to give a reliable estimate of divergence applies to this method also, we shall not go into details.

The relationship between vertical velocity and divergence

Matter is neither created nor destroyed. Consequently, if a lump of fluid experiences divergence, i.e. its horizontally projected area is increasing, then its vertical extent must be decreasing; conversely, if it experiences convergence so that its horizontally projected area is decreasing, then its vertical extent must be increasing. This is illustrated in figure 4.4. Of course, an alternative is that mass accumulates (or is depleted) in some regions so that the density increases (or decreases), but the horizontal variations in density are of minor importance for synoptic-scale motion.

If the earth's surface is horizontal the vertical velocity must be zero right at the ground so that upward motion in the mid-troposphere is associated with

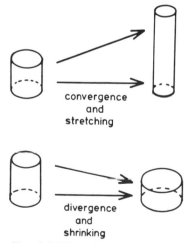

Fig. 4.4 The relationship between divergence and vertical velocities. Converging columns of air must stretch and diverging columns must shrink.

convergence in the lower troposphere. Moreover, because vertical velocities are small in the stratosphere, it must also be associated with divergence in the upper troposphere. Likewise, downward motion in the mid-troposphere has divergence below and convergence aloft. The situation is sketched in figure 4.5.

It is possible to be quantitative about the relationship between divergence and the vertical velocities. Here, it is convenient to express the vertical velocity in terms of the rates of change of pressure experienced by the air particles, because this leads to a particularly simple formula. As air particles rise, they find less air above them and therefore their pressure falls. Conversely, if they sink, their pressure rises. The rate of change of pressure experienced by an air particle (or 'pressure velocity') is usually given the symbol ω. Practical units for ω may be millibars per hour, but when used in equations consistent units must be employed which will be Newtons per square metre ($N\ m^{-2}$) for pressure and seconds for time, giving $N\ m^{-2}\ s^{-1}$ for ω. Pressure velocities may be somewhat unfamiliar, but they can be related with reasonable accuracy to the usual vertical velocities by the approximate formula $\omega = -\rho g w$. Here w is the conventional vertical velocity in $m\ s^{-1}$, ρ the density in $kg\ m^{-3}$ and g is the acceleration due to gravity in $m\ s^{-2}$. Note that if the particle is moving upwards, it will be moving towards lower pressure so ω will be negative. Now consider two points, one above the other. If ω_1 and ω_2 are the 'pressure velocities' at the upper and lower points and if p_1 and p_2 are the pressures of those points then

$$\text{Horizontal divergence} = -\frac{\omega_1 - \omega_2}{p_1 - p_2}. \tag{4.2}$$

This relationship is known as the continuity equation since it is really a statement that matter is neither created nor destroyed. This is further discussed by Panofsky (1981).

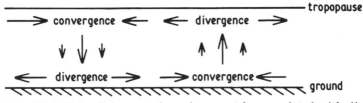

Fig. 4.5 Mid-level downward motion must be associated with divergence below and, since stratospheric vertical velocities are small, convergence aloft. Mid-level upward motion likewise has convergence below and divergence aloft.

Vorticity

Let us turn now to vorticity, which is a measure of the 'spin' of fluid particles. Once more we need to caution the reader that, for the fluid dynamicist, 'spin' is usually a three-dimensional affair. This may be followed up by reading Scorer (1957). However, on the synoptic scale, horizontal motions are in some ways dominant, and we shall consequently confine our attention to spin about a vertical axis produced by the horizontal part of the motion. Even so, lumps of fluid deform in quite complicated ways so that 'spin' is not such a straightforward concept as it is for solid objects.

A good way to picture what is involved is to imagine two short lines of marked fluid particles initially perpendicular at AB and BC as in figure 4.6. After a certain time, say t seconds, the motion will have carried the fluid from A to the new position A', B to B' and C to C'. If μ is the angle which A'B' makes with AB and ν the angle which B'C' makes with BC then μ/t (= α, say) and ν/t (= β, say) are the average rates of rotation of the marked lines during the t seconds.

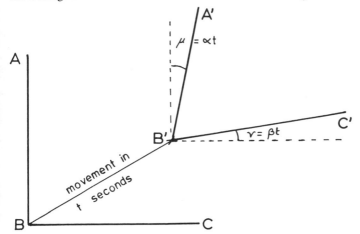

Fig. 4.6 AB and BC are two lines of smoke at their moment of release. After t seconds they have been carried to positions A'B' and B'C' respectively.

It is natural to call the mean of these two rotation rates the 'spin' of the fluid at B. It is conventional to take the anticlockwise (cyclonic) direction as positive so β has been measured in the positive sense and α in the negative. Taking this into account gives

('spin' of fluid at B (relative to Earth)) = $\frac{1}{2}(\beta - \alpha)$.

As in the case of divergence, the spin is strictly meaningful only for a given point and instant so we again need to find the value of β and α with t, AB and BC as small as possible. When this is done we shall call $\beta - \alpha$ the 'vorticity'*. So

'vorticity' = 2 x 'spin'.

A frequently used symbol for vorticity is ζ. So far, spin has been measured relative to the Earth. Sometimes, however, we shall need it measured with respect to absolute space. Consequently, to avoid confusion we shall use ζ_r for the vorticity relative to the Earth and ζ_a for the vorticity relative to absolute space.

---

* α and β should be in radians per second, π radians = 180 degrees. The way the radian is defined means that it does not have dimensional units so the units of α, β are s^{-1}.

Readers familiar with partial derivatives can probably convince themselves that $\alpha = (\partial u/\partial y)$ and $\beta = (\partial v/\partial x)$, where x, y are distances measured parallel to BA and BC respectively and u and v are components of the horizontal velocity in the respective directions. Thus,

$$\zeta_r = \frac{\partial v}{\partial x} - \frac{\partial u}{\partial y}.$$

The difference between the spin about a vertical axis relative to absolute space and the spin relative to Earth is obviously just the component of the Earth's spin about the vertical which is shown in figure 4.7 to be $\Omega \sin \theta$. The difference between the relative and absolute vorticities will be twice this, namely $2\Omega \sin \theta$. This is the familiar *Coriolis parameter* which is usually given the symbol f. Hence

$$\zeta_a = f + \zeta_r.$$

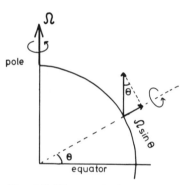

Fig. 4.7 The vertical component of the Earth's rotation is $\Omega \sin \theta$, where θ is the latitude and Ω the rate of rotation about the polar axis.

Shear and curvature

Most people will find some difficulty in mentally computing $(\partial v/\partial x) - (\partial u/\partial y)$ by looking at a weather map. Fortunately, there is an alternative expression for the relative vorticity which can be pictured more readily. Figure 4.8 illustrates what is involved. The relative vorticity comprises two terms, curvature and shear.

Thus,

$$\zeta_r = \frac{V}{R} + \frac{\Delta V}{\Delta R}. \tag{4.3}$$

The curvature term is V/R, namely the windspeed, V, divided by the radius of curvature, R, of the streamline. To find the radius of curvature of the streamline at P we take a short segment of the streamline through P and imagine this as part of a circle, the radius of which is radius of curvature. There is a sign convention that cyclonic curvatures have positive, and anticyclonic, negative radii of curvature (vice versa in Southern Hemisphere).

The shear term, $\Delta V/\Delta R$, is the rate at which the wind speed increases in the direction perpendicularly to the right of the streamline — for instance 1.2 metres per second per 100 km (= 1.2×10^{-5} s^{-1}). If this is a positive number the shear is 'cyclonic' but if it is negative, that is, if the wind speed *decreases* to the right, the shear is anticyclonic. Figure 4.9 illustrates some combinations of cyclonic and anticyclonic shears and curvatures.

It is comparatively easy to visualize whether the curvature and shear are positive or negative and to gain an 'eyeball' impression of whether they are large or small. Nonetheless, if they have opposite signs, it may be difficult to guess which is dominant.

Measuring the vorticity

The previous section may have already suggested to the reader how he may measure vorticity for himself. Estimating the shear is pretty straightforward on the basis of the geostrophic winds, or better from an analysis of the real wind speeds. The main difficulty arises in measuring the radius of curvature. A convenient way is to draw circles of known radius on a piece of transparent material and see which one matches the streamline curvature best at the point in question. Although this requires a certain subjective judgement, it is simpler than the alternative method of finding the centre of curvature as the point of intersection of two lines drawn perpendicular to the streamline a short distance on either side of the studied spot.

Using the same notation as in the section on measuring divergence (see figure 4.3) it is fairly obvious that our expression for vorticity is equivalent to

$$\zeta_r = \frac{v_C - v_G}{2d} - \frac{u_A - u_E}{2d}.$$

This time there is no awkward cancellation if geostrophic winds are employed. On the contrary, a convenient expression for the relative vorticity at O in figure 4.3 based on the geostrophic approximation is:

$$\zeta_r = \{p_O + p_C + p_E + p_G - 4p_O\}/(\rho f d^2). \tag{4.4}$$

To see this we note that v at the midpoint of OC is $(p_C - p_O)/(\rho f d)$ and that v at the midpoint of GO is $(p_O - p_G)/(\rho f d)$ so that an estimate of $\partial v/\partial x$ at O is $(p_C + p_G - 2p_O)/(\rho f d)$. Finding $\partial u/\partial y$ in like manner leads to the result. We have assumed ρ and f constant which is reasonable in comparison with the variations in p.

The vorticity theorem

The principles governing the vorticity change of the air as it flows from place to place are familiar to all as it is used by ice skaters and these may be watched on television almost any evening when no major horse jumping event is taking place. The skater gets himself spinning with arms extended sideways and possibly

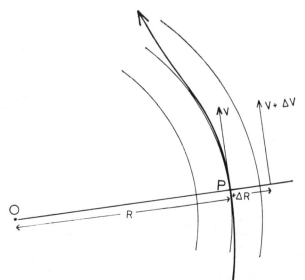

Fig. 4.8 The meaning of the terms in Equation 4.3. The vorticity is to be measured at the point P. The segment of streamline near P can be regarded as part of a circle centre O, radius R. The windspeed is V at P and increases by an amount ΔV in a distance ΔR to the right of P.

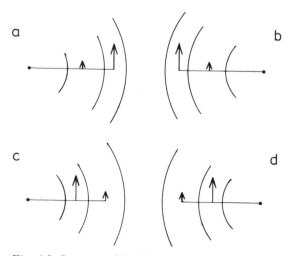

Fig. 4.9 Some combinations of curvature and shear.
    (a) Curvature and shear both positive (cyclonic).
    (b) Curvature and shear both negative (anticyclonic).
    (c) Positive curvature and negative shear.
    (d) Negative curvature and positive shear.

crouching with one leg stretched out. The arms and legs are then drawn in and the body straightened so that as nearly as possible the whole body is in a vertical line. This has the effect of markedly increasing the spin rate. I recently saw an example of this where the skater spun so fast that he had difficulty stopping. It is instructive to try the same thing for yourself standing on a swivel chair but some care is needed to avoid ending in an undignified heap on the floor. The physical principle at work is the conservation of angular momentum. This states that if there is no net torque acting on a rotating object then the product of the angular velocity (spin rate) and the 'moment of inertia' is constant. The moment of inertia is a measure of how the mass is distributed: if the mass is a long way from the axis of rotation the moment of inertia is large, if it is close to the axis, the moment of inertia is small. When the skater draws himself into a line he brings all his mass close to the axis of rotation so that his moment of inertia is reduced. Thus, to keep the product of angular velocity and moment of inertia constant, the angular velocity has to increase. To slow down the arms are spread out sideways again and perhaps a leg extended. This increases the moment of inertia so the angular velocity decreases to keep the product constant.

The same process happens in the atmosphere; as air spreads out horizontally the spin rate decreases. Conversely, if the air contracts in the horizontal, the spin rate increases. We anticipate therefore that the rate of change of vorticity is related somehow to the divergence. The relationship turns out to be very straightforward because it is not difficult to show that the fractional rate of change of the moment of inertia about a vertical axis of small, circular fluid elements is div $\mathbf{v}_h$ while the fractional rate of change of the angular velocity is obviously $(1/\zeta_a) \{(d/dt)\zeta_a\}$. If the product of moment of inertia and angular velocity remains constant, these fractional rates of change must be equal and opposite giving

$$\frac{d}{dt}\zeta_a = -\zeta_a \text{ div } \mathbf{v}_h. \tag{4.5}$$

This is our main equation for the rate of change of vorticity: *the rate of change of absolute vorticity of particular portions of fluid is equal to minus the absolute vorticity multiplied by the divergence.* Remembering that divergence is related to vertical velocities the physical meaning is as in figure 4.10.

We now have all the theoretical ideas which we need. In the remainder of this chapter we use them to understand some familiar synoptic features; but first some comments on the validity of the vorticity equation are in order.

In general, equation 4.5 is true for motion on the synoptic scale outside the planetary boundary layer, but it is really an approximation and there should be other terms on the right-hand side, albeit small. For instance, horizontal gradients of vertical velocity can change spin about a horizontal axis into spin about a vertical axis. Roughly speaking, this term is expected to be small for those motion systems where the relative vorticity, ζ_r, is small compared with the Coriolis parameter, f. This is usually, but not always, the case in mid- and high latitudes. There is another mechanism for creating vorticity which has not been included

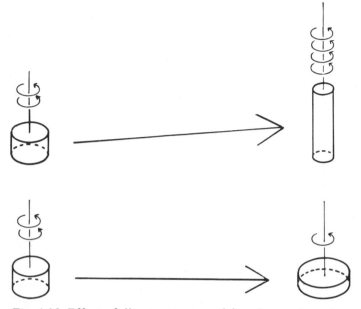

Fig. 4.10 Effect of divergence on vorticity. Converging columns have increasing absolute vorticity, diverging columns have decreasing absolute vorticity.

in equation 4.5. This operates when surfaces of constant pressure do not coincide with surfaces of constant temperature. This too is usually unimportant for mid- and high-latitude synoptic-scale motion and will not be considered further. Scorer (1958) contains many fascinating examples of both effects. Moreover, the effects of friction have also not been included in the equation about which more will be said below.

Rossby waves

We are now able to put our insight to use in explaining some familiar phenomena. The simplest case to begin with is one of purely horizontal motion. Zero vertical velocity implies no divergence or convergence as we have already seen, so the rate of change of absolute vorticity of air packets must be zero. That is, their absolute vorticity, ζ_a, is constant. Let us fix attention on one particle and assume that its absolute vorticity is the same as the Earth's vorticity at a particular latitude. We shall call this value f_0. Thus for the chosen particle $\zeta_a = f + \zeta_r = f_0$. Now f is proportional to the sine of the latitude (figure 4.7) so that, south of the reference latitude, f is less than f_0. Hence, to keep $\zeta_a = f_0$ the relative vorticity must be positive there. Likewise, if the particle is to the north of the reference latitude, its relative vorticity must be negative. Given a sufficiently broad airstream so that the shear term of equation 4.3 is small, it follows that, when the particle is north of the reference latitude, the streamline curvature must be anticyclonic, and when it is south of it the curvature must be cyclonic. This gives the possibility of a steady, stationary train of waves in a westerly current as sketched in figure 4.11. These are called Rossby waves. A more complete mathematical treatment shows that for the waves to be steady, there has to be a particular windspeed depending on how many troughs and ridges occur around a latitude circle. An average westerly flow of about 15 m s^{-1} gives a stationary wave train with three troughs and three ridges, whilst just over twice this gives a stationary wave train with two troughs and two ridges. If charts from the upper and middle troposphere are averaged for about a month to remove transient phenomena like travelling highs and lows, the resulting map does indeed show a rather broad westerly flow disturbed by large-scale meanders having two or three more or less well-defined troughs. An example is given in figure 4.12, which shows the average flow for January in the mid troposphere at a height of about 5 km. There are well-marked troughs at $70°W$ and at $170°E$ with a faint suggestion of another near $70°E$. Mid-latitude wind speeds lie between 12 and 30 m s^{-1}. The configuration agrees fairly well with what we expect according to our theory, though many details remain to be explained.

When Rossby wave theory is developed in more detail it predicts that, if the basic westerly flow is less than the critical speed, the waves drift westwards, while, if it is more, they drift eastwards. In this non-stationary case the particle trajectories are not streamlines and arguments like those portrayed in figure 4.11 need considerable circumspection. Such arguments can be used, however, to show that stationary Rossby waves cannot exist in flow which is basically from east to west: the reader may like to try this for himself.

In practice, the basic westerly wind is not the critical velocity for stationary waves and yet, since the waves are initially generated by mechanisms fixed on the Earth (we shall consider one of them below), the waves have to be stationary

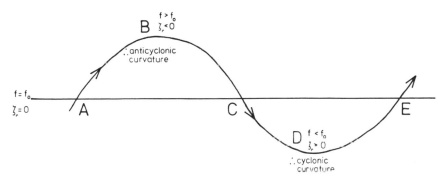

Fig. 4.11 The simple Rossby wave. The particle moves north-east from A where $f = f_0$ and $\zeta_r = 0$, so that $\zeta_a = f_0$. When it reaches B, $f > f_0$, so that to keep ζ_a constant, $\zeta_r < 0$. This gives anticyclonic curvature. Consequently, the flow turns towards C. Upon overshooting to D the particle here needs $\zeta_r > 0$, i.e. cyclonic curvature. Thus, the trajectory returns northwards to E.

45

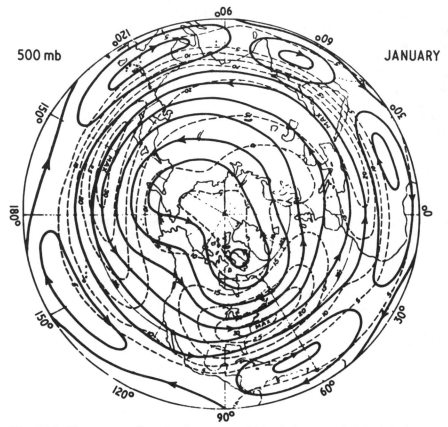

Fig. 4.12 The average flow for January at 500 mb (approx. height 5 km). Solid lines are streamlines, broken lines are lines of equal wind speed labelled in m s^{-1}. (Taken from WMO No. 111, TP49 after Mintz and Dean, 'The observed mean field of motion of the atmosphere', *Geophysical Research Papers No. 17.*)

nonetheless. To achieve this, vertical motions are necessary as well as horizontal. This in turn imposes a particular vertical structure on the wave. In the winter, the planetary-scale stationary waves may be traced all the way up to the mesosphere and beyond, to heights of 90 km. They provide an important mechanism through which the upper and lower layers of the atmosphere may interact with each other and are therefore the subject of much current research.

The lee trough

When a stable airstream flows across a substantial mountain range a trough or surface cyclone often develops on the lee side. A consideration of the vorticity shows why this should be. Air near the surface is constrained to follow the ground shape, rising up the windward side and subsiding down the lee. The air aloft, however, attempts to flow horizontally, the more so the more stable the airstream is. Accordingly, air particles in the lower part of the atmosphere shrink in the vertical as they pass over the windward slope and then are stretched again as they descend in the lee (see figure 4.13). Hence, the absolute vorticity will be reduced while the air ascends the windward slope and subsequently returned to the initial value as it descends the lee. If, for simplicity, we assume that the air arrives at the forward edge of the mountain with zero relative vorticity, the

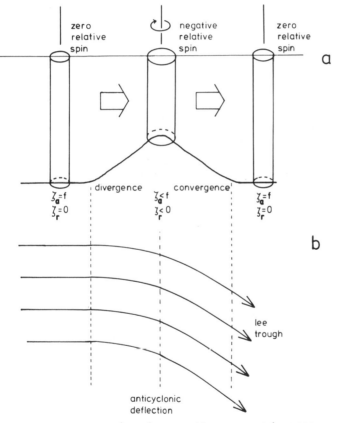

Fig. 4.13 Deflection of an air current by a mountain range.
(a) Side view
(b) Plan view
The flow arrives at the forward edge of the range with zero relative vorticity. As the ridge is ascended, the air columns are shortened in the vertical and so diverge in the horizontal. Their absolute vorticity is therefore reduced below f giving negative relative vorticity. Accordingly, the flow is deflected anti-cyclonically. Possible variations in f have been neglected in this diagram.

relative vorticity will be negative (anticyclonic) all the time the air is over the mountain. Again, for simplicity, we shall assume that both the current and the mountain range are broad in the cross-flow direction so that the horizontal wind shear can be neglected. The negative relative vorticity over the mountain then implies anticyclonic curvature, i.e. the flow is deflected to the right. Since low pressure lies to the left of the stream line, the pressure is lower on the lee side of the mountain than on the windward side, giving a 'lee trough'. A striking example of this deflection to the right produced by the Alps is shown in figure 4.14.

The lee trough formed to the east of the Rockies is one of the important mechanisms for initiating the north—south meanders of the Rossby waves.

Potential vorticity

Because air columns which converge increase both their vertical extent and their absolute vorticity and those which diverge decrease both, we may guess that there is a quantity like the absolute vorticity divided by the length of an air column which remains constant. This hunch proves to be correct, except that the relevant divisor is not the length of the air column but rather the pressure difference between the top and bottom of the column. Thus, if the pressure at the top of the column is instantaneously p_1, and the pressure at the bottom is p_2, then as the particle moves about, the ratio $\zeta_a/(p_2 - p_1)$ stays constant. This ratio (strictly speaking something very like it) is called the 'potential vorticity'.

To illustrate the usefulness of this idea, let us put some typical magnitudes into the idealized case shown in figure 4.13 to show that we can indeed expect curvatures of the magnitude found in figure 4.14. Suppose the height of the mountain in figure 4.13 is 3000 m and the surrounding plain is at sea-level. The usual distribution of pressure with height implies that the pressure at the bottom of the air column (p_2) is about 1000 mb on the plain but about 700 mb when the mountain crest is reached. We cannot easily say at what height the flow is unaffected by the mountain but, for the sake of argument, we take it to be the 300 mb level (something over 9 km), so that $p_1 = 300$ mb wherever the column is found. Assuming $\zeta_r = 0$ over the plain, and using ζ_m for the *relative* vorticity over the mountain, the constancy of potential vorticity gives:

$$(f + \zeta_m)/(700 - 300) = (f + 0)/(1000 - 300),$$

so that

$$\zeta_m = -\frac{3f}{7}.$$

Now neglecting the shear, equation 4.3 states

$$\zeta_m = \frac{V}{R},$$

where R is the radius of curvature of the flow at the mountain crest. Combining the last two equations gives

$$R = -\frac{7V}{3f}.$$

At $45°N$, $f = 10^{-4}$ s^{-1}, so if we have the reasonable windspeed $V = 10$ m s^{-1}, the radius of curvature turns out to be 230 km. Presumably, the shrinking of the column happens more at the bottom than at the top, so we may expect smaller radii of curvature near the surface but large ones aloft. The curvatures seen in figure 4.14 are readily understandable on this basis.

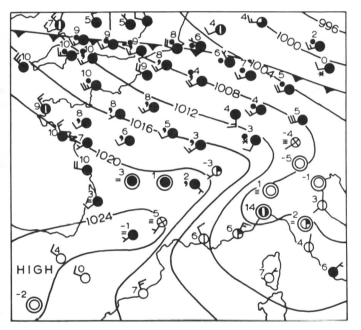

Fig. 4.14 Anticyclonic deflection of an air current by the Alps. From *Daily Weather Report* 0600 GMT, 6 December 1967.

To prove the constancy of potential vorticity it is necessary only to realize that the 'pressure velocity', ω, used in equation 4.2 is by definition dp/dt, so that the equation may be rewritten

$$\text{div } \mathbf{v}_h = - \left\{ \frac{d}{dt}(p_2 - p_1) \right\} / (p_2 - p_1).$$

If this is substituted into equation 4.5, the result follows after very little manipulation.

The heat low

During the summer months air over continental land masses becomes very hot. If the heated region is of fairly small scale, as for example in the case of Spain, surrounded by cooler air over the sea, a heat low develops. Figure 4.15 shows a typical situation.

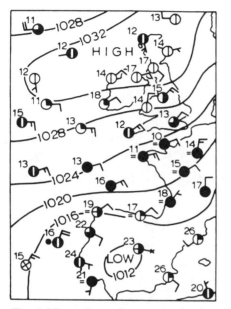

Fig. 4.15 A heat low over Spain. From *Daily Weather Report* 1200 GMT, 29 May 1963.

Consider such an isolated air mass which, for simplicity, we will take to be initially motionless. Figure 4.16 shows what happens when heat is applied. The heated air is buoyant compared with its surroundings and therefore rises. This raises the pressure at B and since there is initially no motion to give a Coriolis force, the pressure gradient causes the air aloft to flow out horizontally. This divergence lowers the absolute vorticity at B which therefore gives an anticyclone there. The divergence aloft means that there is less mass in the column so that the surface pressure becomes lower than the surroundings. As before, the initial condition of no motion allows this pressure gradient to accelerate air inwards giving convergence at A which increases the absolute vorticity and gives rise to a surface cyclone there. The convergence and divergence stop when the horizontal velocities in the surface cyclone and upper anticyclone are enough for their Coriolis forces to balance the pressure-gradient forces.

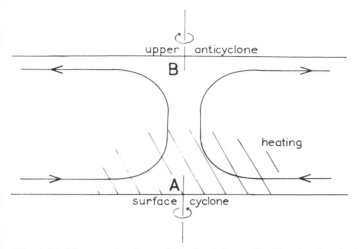

Fig. 4.16 The mechanism of the heat low. Heating in the shaded region causes air to rise in middle levels. Here it is shown diverging at *B* below a stable layer (for example the tropopause). The divergence at *B* lowers the absolute vorticity and produces an anticyclone there. At *A*, however, the air must converge and a surface heat low develops.

Mid-latitude cyclones

The vorticity equation can be very usefully applied to the mid-latitude cyclone and anticyclone. In what follows, we shall be considering air in the bottom few kilometres unless we state otherwise.

Near the centre of the cyclone, relative vorticity is positive, whereas near the centre of the anticyclone it is negative. Accordingly, as air flows into a cyclone, its absolute vorticity is increasing. It must therefore be subject to convergence, implying middle-level ascent. Two consequences may be anticipated. The first is that there will be a widespread cloud area, and perhaps rain associated with the region where the air flows into the centre. The second is that deep convection is more likely to be found in regions of high absolute vorticity than elsewhere, since the process of stretching the air columns needed to produce the large vorticities moves the potential temperature surfaces further apart in the vertical. In consequence, penetrative convection capable of reaching a given potential temperature level will attain greater heights than in low vorticity regions where the potential temperature surfaces are closer together. The situation is reversed for the anticyclone: as air enters the anticyclone it must be subjected to divergence and middle-level descent. The sky should therefore become free of large-scale cloud, and penetrative convection in the low (absolute) vorticity air should remain shallow as the potential temperature surfaces will have been brought close together.

In some respects, the surface weather chart gives a misleading impression of where the regions of vorticity creation and destruction are. This is because friction causes the surface winds to be backed against the isobars (see Panofsky, 1981). In consequence, all regions of cyclonic relative vorticity have convergence in the friction layer and all regions of anticyclonic relative vorticity have divergence. However, it would be a mistake to assume that this appreciably alters the vorticity, for there is another effect of friction which we have neglected in equation 4.5, namely that friction tends to slow down the motion, thus attempting to bring the relative vorticity to zero. To a first approximation the two effects cancel; the rate at which the frictionally induced convergence increases the relative vorticity in cyclones is almost exactly the rate at which friction reduces it. A similar statement applies in anticyclones. What we must be concerned with is the divergence or convergence in the free air above the friction layer. We shall infer what this is, not by looking at anemometer level winds, but by observing the changes of vorticity experienced by the air particles outside the friction layer.

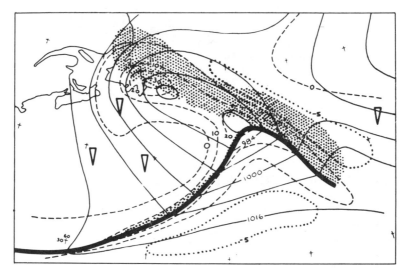

Fig. 4.17 Surface chart for 0600 GMT, 17 December 1958. Solid lines are isobars at 8 mb spacing. Pecked lines show relative vorticity in 10^{-5} s^{-1}. Stippled area shows continuous rain reported.

Figure 4.17 shows a depression crossing the Atlantic in December 1958. The dotted lines show the relative vorticity in units of 10^{-5} s^{-1}. This has been worked out from the pressures using equation 4.4 with $d = 200$ km. It therefore overestimates the vorticity of the air at anemometer level because of friction. It should, however, be reasonably representative of the vorticity at a height of, say, 500 m to 1 km.

Most of the high values of vorticity (up to about $4f$ in places) are associated with the low pressure centre, but there are bands of high vorticity spreading along the fronts. The width of these bands along the front is exaggerated by the computational technique. Really, the frontal structure should show far higher values of vorticity occupying a much narrower region, but equation 4.4 effectively finds the vorticity averaged over a 400 km square. There is a considerable region of anticyclonic relative vorticity in the cold air sweeping southwards behind the centre, even though the streamline curvature is cyclonic.

To relate the vertical velocities to vorticity changes we need to know how the vorticity of individual portions of air changes. This is a little difficult to visualize from figure 4.17 because although the velocity of the air parcels relative to the surface are readily seen from the pressure pattern, the system moves as well, in this case at 25 m s^{-1} towards the east-north-east. It is more convenient to observe the behaviour in a frame of reference which moves with the system. In this frame of reference the complications of the translation of the system are removed. Accordingly in figure 4.18 a pressure pattern which corresponds to the movement of the system has been subtracted, and geostrophic winds implied by the isobars in that figure are approximations to winds as they would be experienced by an observer moving east-north-eastward at 25 m s^{-1} so as to stay in the same spot relative to the system. But, because of the comparatively slow development of the system, the streamlines of figure 4.18 are also trajectories: where the flow is towards areas of high vorticity the air columns must be subject to convergence and where the flow is towards areas of low vorticity the air columns must be subject to divergence. It follows that where the flow is towards high

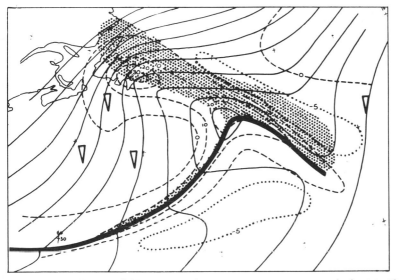

Fig. 4.18 As Figure 4.17, except that a pressure pattern equivalent to the movement of the low has been subtracted (25 m s^{-1} to east-north-east). The 'isobars' in this figure therefore give approximate streamlines relative to the system and show how the air is entering and leaving the high vorticity region.

values of vorticity there should be ascending motion, and where the flow is towards low vorticity there should be sinking motion. The agreement is quite satisfactory since the stippled rain area, which was analysed from the observations before ever the vorticity or relative winds were computed, lies well in the ascending air. The descending air is in the rear of the cyclone where such cloudiness as exists is usually associated with deep shower convection formed as the cold air sweeps south over progressively warmer sea, the skies between these showers being clear and the air relatively dry.

These have been just a few examples of how thinking about the vorticity of the air motion helps us to piece together the separate features of the atmosphere's behaviour into a coherent whole. There are others which we could have mentioned, such as the weather pattern in an easterly wave, or what happens in polar lows, or why cyclogenesis is more likely to occur under the right entrance region of a jet stream than the left, but perhaps enough has already been said to encourage the reader to interpret his weather map in a new and instructive light.

References

Jeffreys, H. and Jeffreys, B. S. (1946) *Methods in Mathematical Physics*, Cambridge, Cambridge University Press.
Panofsky, H. A. (1981) 'Atmospheric hydrodynamics', this volume 8–20.
Scorer, R. S. (1957) 'Vorticity', *Weather*, 12, 72–83.
Scorer, R. S. (1958) *Natural Aerodynamics*, New York, Pergamon Press.

5
Practical analysis of dynamical and kinematic structure: principles, practice and errors

M. A. PEDDER
University of Reading

The dynamical meteorologist is frequently interested in the relationship between various properties of the atmosphere which cannot be measured directly, but which are related to observable quantities by differential equations with respect to space or time. Typical properties of this type, which have already been discussed elsewhere in this volume, are the pressure-gradient force and the corresponding geostrophic wind vector (see Panofsky, 1981) and the horizontal wind divergence and relative vorticity (see Harwood, 1981). In order to test dynamical theories, to establish which dynamical or physical processes are most important during a particular sequence of events, to prepare initial fields for numerical prediction, and to relate the dynamical and kinematic structure of the atmosphere to observed weather phenomena, we must be able to estimate these properties from routine observations of pressure, wind, temperature, etc.. This chapter is concerned with how such estimations can be made, and to what extent their accuracy is limited by errors of observation and sampling.

General principles

Most routine observations of pressure, temperature, wind, and other meteorological variables are recorded at fixed points in space (as defined by the map locations of surface and upper air stations) and time (as, for example, defined by the synoptic hours 0000, 0600, 1200 and 1800 GMT). However, in order to obtain estimates of dynamical and kinematic variables from these *discrete* data it is, in principle, necessary to produce a more or less continuous representation of each observable quantity in space and time. For simplicity, I shall consider mainly the problem of representing variations in horizontal map space at a fixed height (or pressure) and fixed time, as represented, for example, by the mean-sea-level surface pressure map or the map of geopotential height (h) of a constant pressure surface. In the case of the geopotential height map, the analysed chart can be considered as representing a surface in h, x, y space as sketched in figure 5.1, where x and y define the map co-ordinates in the horizontal corresponding

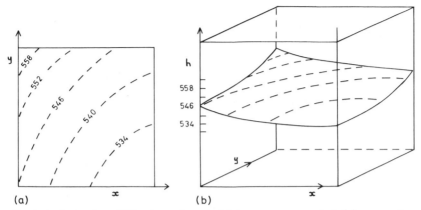

Fig. 5.1 Part of a 500 mb surface height map represented (a) as contours of height (h) in x, y map space, and (b) as a surface in h, x, y space.

to a particular height value, h. If the shape of this surface is known, then estimates of important derivative properties such as pressure-gradient force, geostrophic wind and the relative vorticity of the geostrophic wind at any point on the map can be calculated from the local slopes, or higher space derivatives, of this surface with respect to the x and y directions. In practice we do not need to represent the continuous variation of h in x, y space, since these estimations can be related more directly to observations by one of the two following methods: (i) By *interpolating* the observations from irregularly spaced stations onto a regular array of grid points, and using a finite differencing technique to estimate the required gradient properties (see Harwood, 1981; Gadd, 1981); or (2) By *fitting* the observations to some function of x and y which can then be differentiated analytically for any chosen pair of x, y values.

Of the two methods, the first is most commonly applied to large-scale map analysis. The interpolation can be carried out either by using a traditional 'hand'-analysed chart on which is superimposed the network of grid points, or numerically by means of an equation relating the grid point values to the observations directly, the latter being termed an 'objective' analysis and the former a 'subjective' analysis. The second method is often preferred for the estimation of quantities derived from a relatively small number of observations, and in this day of the pocket calculator, can be made a rather quicker method than the subjective analysis technique, apart from providing more reproducible (though not necessarily more accurate) results. In the next section, I shall describe what is perhaps the simplest example of this objective analysis method, and in Chapter 6 describe some applications.

In relation to the objective analysis of dynamical or kinematic structure, it would appear at first sight as though methods (1) and (2) are quite dissimilar. However, it can be shown that, in both cases, the variable $F(q)$ to be estimated from observations of the quantity q may generally be expressed in the form

$$\hat{F}(q) = w_1 q_1' + w_2 q_2' + \cdots w_i q_i' \cdots + w_m q_m' = \sum_{i=1}^{m} w_i q_i', \qquad (5.1)$$

where the symbol $\frown$ denotes an estimation, q_i' is an observed value of q at station i, and w_i is a number representing a *weight*, given to the ith observation, which depends only on the position of the corresponding ith station in the array of m stations actually used for the estimation. This deceptively simple-looking equation defines a *linear* relation between $\hat{F}(q)$ and $q_1', q_2' \ldots q_m'$, and even quite sophisticated objective analysis schemes can usually be reduced to a form similar to equation 5.1. The main problem of practical analysis lies in the choice of weights $w_1, w_2 \ldots w_m$, since there is an apparently infinite set of possible combinations to choose from, and any one combination will generally result in a different value of $\hat{F}(q)$ than that given by a different combination. This means that no single analysis scheme can be guaranteed to give the best possible estimation of $F(q)$ in any one situation, and that no two different analysis schemes are likely to give exactly the same result. The underlying reason for this uncertainty in what is the 'best' choice for the weights, $w_1, w_2 \ldots w_m$, is that, in meteorological analysis, we have no *a priori* knowledge concerning the precise shape of the surface being analysed, so that while every different analysis scheme is effectively producing a different 'guess' for the shape of the surface, there is no way of knowing which is the best representation for the problem in hand.

The 'least-squares-plane' method

Many important dynamical and kinematic quantities can be expressed in terms of the local first-order space derivatives of an observable quantity q, these being $\partial q/\partial x$ and $\partial q/\partial y$ in horizontal map space. For example, the geostrophic wind components u_g and v_g at a particular pressure level are related to $\partial h/\partial x$ and $\partial h/\partial y$ by

$$u_g = -\frac{g}{f}\frac{\partial h}{\partial x}, \qquad v_g = \frac{g}{f}\frac{\partial h}{\partial y}. \qquad (5.2)$$

A simple method for estimating $\partial h/\partial x$ and $\partial h/\partial y$ from height observations, which does not require any interpolation onto grid points in map space, is to fit a 'plane surface' to the data in h, x, y space, as illustrated in figure 5.2. Mathematically, this plane is represented by the formula

$$h(x, y) = a_0 + a_1 x + a_2 y, \qquad (5.3)$$

where a_0, a_1 and a_2 are constants which can be estimated from the data. From equation 5.3 it is easy to show that $a_1 = \partial h/\partial x$ and $a_2 = \partial h/\partial y$ at all points on the surface, giving us the derivative quantities as required, for example, to solve equation 5.2.

Of course, since the $h(x, y)$ surface is not actually a plane surface, equation 5.3

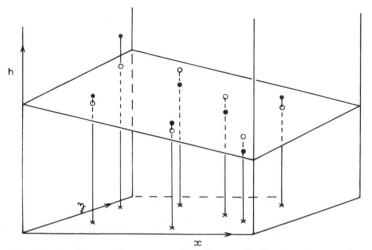

Fig. 5.2 A plane surface approximation to the local variation of the geopotential
height (h) of a pressure surface in h, x, y space. ● corresponds to the true
value of h, and ○ to the plane surface approximation of h at the position
marked X.

will represent only an approximation to the true surface, though it will be a
reasonable approximation for sufficiently small areas of map space. Where the
area of analysis is not 'small', we may nevertheless consider a_1 and a_2 as representing the *mean slope* components of the true $h(x, y)$ surface within the area of
interest, which is a valid and useful concept in practical analysis, since we are
often not interested in the 'fine detail' of $h(x, y)$, but only in its local 'trend'.

In order to solve for the constants a_0, a_1 and a_2 we need at least three
observations from three different points in map space to give 'three equations in
three unknowns'. For reasons explained later, it is preferable to use more than
three observations, in which case equation 5.3 is said to be 'over-determined'.
Whether we use three or more observations, there is no guarantee that our solution
corresponds to the most representative plane surface for the region of interest
(see following p. 61), so that our solutions for a_0, a_1 and a_2 must always be considered as estimations, written $\hat{a}_0$, $\hat{a}_1$ and $\hat{a}_2$.

In the over-determined case, the solution of interest is normally that which
minimizes the sum of the squared differences between the observations $h'_1, h'_2 \ldots$
h'_m and the values given by equation 5.3 at $x_1, y_1, x_2, y_2, \ldots x_m, y_m$. This
solution is commonly known as the 'least-squares-plane'. For practical purposes,
it is convenient to use x and y values relative to the geometric centre X_0, Y_0 of
the set of stations being used, where $X_0 = (1/m) \Sigma_{i=1}^{m} X_i$ and $Y_0 = (1/m) \Sigma_{i=1}^{m} Y_i$.
In the following, X_i and Y_i will be taken as being spherical curvilinear co-ordinates defined by

$$X_i = \lambda_i R \cos \phi_i$$

and

$$Y_i = R \phi_i, \tag{5.4}$$

ϕ_i, λ_i being latitude and longitude values (in radians) and R the radius of the earth. Defining $x_i = X_i - X_0$ and $y_i = Y_i - Y_0$, the estimated coefficients of the least-squares-plane may then be obtained by solving the following formulae:

$$\hat{a}_0 = \sum_{i=1}^{m} \frac{1}{m} h_i' \ (= \bar{\hat{h}}_i), \tag{5.5a}$$

$$\hat{a}_1 = \sum_{i=1}^{m} \beta_{i,1} h_i' \ (= \partial\hat{h}/\partial x), \tag{5.5b}$$

and

$$\hat{a}_2 = \sum_{i=1}^{m} \beta_{i,2} h_i' \ (= \partial\hat{h}/\partial y), \tag{5.5c}$$

where

$$\beta_{i,1} = \left\{ x_i \sum_{j=1}^{m} y_j^2 - y_i \sum_{j=1}^{m} x_j y_j \right\} \Big/ \left\{ \sum_{j=1}^{m} x_j^2 \sum_{j=1}^{m} y_j^2 - \left(\sum_{j=1}^{m} x_j y_j \right)^2 \right\}, \tag{5.6a}$$

and

$$\beta_{i,2} = \left\{ y_i \sum_{j=1}^{m} x_j^2 - x_i \sum_{j=1}^{m} x_j y_j \right\} \Big/ \left\{ \sum_{j=1}^{m} x_j^2 \sum_{j=1}^{m} y_j^2 - \left(\sum_{j=1}^{m} x_j y_j \right)^2 \right\}. \tag{5.6b}$$

In equations 5.5 and 5.6, suffixes i or j identify a station value of h', x or y, and m is the total number of stations used ($m \geqslant 3$). Notice the similarity between equations 5.5a, 5.5b, 5.5c and equation 5.1, demonstrating that our estimations are linearly related to the observations by the 'weights' $1/m$, $\beta_{i,1}$ and $\beta_{i,2}$. The appearance of these formulae suggests that a considerable amount of computation is required before we arrive at our first estimations of $\partial h/\partial x$ and $\partial h/\partial y$. However, since the weights given by equation 5.6a and 5.6b depend only on position, these need only be calculated on a once-and-for-all basis for a given set of stations.* Thereafter, the calculation of the coefficients $\hat{a}_0$, $\hat{a}_1$ and $\hat{a}_2$ involves only substituting the ordered set of observations h_1', h_2' . . . h_m' into the sum–product formulae 5.5a–5.5c. These summations can be effected rapidly on a simple pocket calculator with one memory.

Having decided that several estimations are to be made using the same set of stations, it is useful to prepare several copies of a blank form of the type shown in figure 5.3 onto which the data and results can be copied. In the author's experience, it then takes about the same amount of time to decode and copy out the data as it does to carry out the summations 5.5a, 5.5b and 5.5c!

Of course, the estimated derivative properties such as $\partial h/\partial x$ and $\partial h/\partial y$ are the same for all values of x and y. However, they may be considered as being

---

* A useful check on the accuracy of these calculations is provided by the property $\Sigma_{i=1}^{m} \beta_{i,1} = \Sigma_{i=1}^{m} \beta_{i,2} = 0$. This result, and the formulae relating the β weights to the position variables, are well known in applied mathematics and statistics. More general solutions of the 'least-squares' problem may be found in textooks on numerical analysis e.g. Dahlquist and Björck (1974).

Form for Planimetric estimations.

7 station array. Array centre: 52.9°N, 3.7°W

Station index i	Station Name	WEIGHTING FACTORS $(10^{-7}\ m^{-1})$		OBSERVATIONS
		$\beta_{i,1}$	$\beta_{i,2}$	$q'_i(\quad)$
1	SHANWELL	0.595	11.510	
2	LONG KESH	-3.911	5.821	
3	AUGHTON	1.097	1.885	
4	HEMSBY	8.453	-1.692	
5	CRAWLEY	6.003	-6.653	
6	CAMBORNE	-2.097	-8.916	
7	VALENTIA	-10.140	-1.955	

$$\left\{\frac{1}{m}\right\}^{\frac{1}{2}} = 0.378 \qquad \frac{1}{7}\sum q_i = \qquad \left\{\frac{1}{m}\right\}^{\frac{1}{2}}\sqrt{\overline{\varepsilon_q^2}} =$$

$$\left\{\sum \beta_{i,1}^2\right\}^{\frac{1}{2}} = 1.52 \times 10^{-6}\ m^{-1} \qquad \sum_{i=1}^{7} \beta_{i,1}\, q'_i = \qquad \left\{\sum \beta_{i,1}^2\right\}^{\frac{1}{2}}\sqrt{\overline{\varepsilon^2}}_q =$$

$$\left\{\sum \beta_{i,2}\right\}^{\frac{1}{2}} = 1.73 \times 10^{-6}\ m^{-1} \qquad \sum_{i=1}^{7} \beta_{i,2}\, q'_i = \qquad \left\{\sum \beta_{i,2}\right\}^{\frac{1}{2}}\sqrt{\overline{\varepsilon^2}}_q =$$

SOLUTION: $\qquad \hat{\bar{q}} = \qquad \pm \qquad ; \quad \frac{\widehat{\partial q}}{\partial x} = \qquad \pm \qquad ; \quad \frac{\widehat{\partial q}}{\partial y} = \qquad \pm$

Fig. 5.3 Example of a blank form used in the least-squares-plane estimation of $\bar{q}$, $\partial q/\partial x$ and $\partial q/\partial y$ from simultaneous observations of q at 7 aerological stations over the British Isles.

estimations 'most appropriate' to the geometrical centre of the station array, i.e. at X_0, Y_0.

Kinematic estimations

So far, I have considered fitting a scalar quantity such as h to a plane surface defined by equation 5.3. However, the same method can also be applied to a vector quantity such as horizontal wind **V**. This must first be decomposed into zonal (u) and meridional (v) components at each station*. The u and v data are then analysed separately (using the same set of weights of course) to provide estimations of $\bar{u}$, $\partial u/\partial x$, $\partial u/\partial y$, $\bar{v}$, $\partial v/\partial x$ and $\partial v/\partial y$. The important derivative properties such as divergence (div **V**) and relative vorticity (ζ_r) are then obtained

---

* $u' = C'x \sin (D' - 180)$, $v' = C'x \cos (D' - 180)$, where C' is reported wind speed in m s^{-1} and D' is reported wind direction in degrees.

from

$$\widehat{\text{div } \mathbf{V}} = \frac{\widehat{\partial u}}{\partial x} + \frac{\widehat{\partial v}}{\partial y} - \frac{\widehat{v}}{R} \tan \bar{\phi}, \qquad (5.7)$$

$$\widehat{\zeta_r} = \frac{\widehat{\partial v}}{\partial x} - \frac{\widehat{\partial u}}{\partial y} + \frac{\widehat{u}}{R} \tan \bar{\phi}, \qquad (5.8)$$

where $\bar{\phi}$ is the latitude corresponding to X_0, Y_0, and R is the earth's radius $(6.36 \times 10^6 \text{ m})$.

With only three stations, div $\mathbf{V}$ and ζ_r can be estimated more directly by a graphical planimetric technique commonly known as 'Bellamy's method' (Bellamy, 1949; Saucier, 1955). However, the pocket calculator estimation based on a plane analysis takes no longer, with practice, than the graphical measurements required for Bellamy's method, while Bellamy's method cannot conveniently be applied to more than three wind observations.

The error problem

It is scientifically bad practice to apply any form of analysis technique without being aware of the magnitude of the probable error on each estimation, and the interpretation of results should always be considered in relation to the probable errors of estimation. For these reasons, it is important to consider the errors typical of these estimations before deciding on what types of measurement might reasonably be attempted using any particular analysis method. In the following, I shall be considering the errors on estimated quantities obtained from the least-squares-plane analysis, though similar error estimations can be made for other types of analysis schemes.

There are really three sources of error in the above types of estimation, due to (1) random observing error; (2) sampling error; and (3) analysis model error*. Error types 2 and 3 I shall be considering only qualitatively, since a quantitative analysis of these effects is mathematically or statistically involved.

Random observing errors

The influence of these errors on the accuracy of an analysis can be estimated in a relatively straightforward way from the weights used in the analysis, provided that the error typical of the observations is known. Of course, for any one simultaneous set of observations we have no way of knowing the observing error on each observation (otherwise we would obviously allow for it before starting our analysis!).

Suppose, however, for the moment that we *could* actually measure the observing error (ϵ) associated with each observation q'. For most types of observations this type of error is such that both positive and negative values are equally likely,

---

* The combined effect of all these sources of error is commonly known as the 'interpolation error'.

and the average value of ϵ $(\bar{\epsilon})$ for a very large number of observations would then be found to be zero. However, the average square of ϵ $(\overline{\epsilon^2})$ would *not* be zero, since both the positive and the negative values of ϵ correspond to positive values of ϵ^2. The quantity $\overline{\epsilon^2}$ is generally known as the *error variance*. The approximate magnitude of this quantity can often be estimated from the technical characteristics of the observing system used to measure the quantity with which ϵ is associated. Thus, the error variance of 500 mb geopotential height (expressed in geopotential metres (gpm)), as estimated using most types of radiosonde system, is reckoned to be about 150 $(gpm)^2$. The square-root of the error variance $(\overline{\epsilon^2})^{1/2}$ is often termed the 'root-mean-square (rms) error'; in the case of the 500 mb geopotential height measurement it is about 12 gpm. The statistical properties of this type of error are such that there is a roughly 67 per cent chance that the magnitude of ϵ on a single observation is less than $(\overline{\epsilon^2})^{1/2}$ and roughly 96 per cent chance that it is less than $2(\overline{\epsilon^2})^{1/2}$. The degree of confidence which can be placed on a single observation may thus be expressed quantitatively in terms of its *rms* error, even though the *true* error is not known.

Suppose we now use a relation of the form of equation 5.1 to estimate a quantity $F(q)$ from m observations of a quantity q which has an rms error known to be $(\overline{\epsilon_q^2})^{1/2}$. It can then be shown that the rms error on the estimation $\hat{F}(q)$ $(e_{\hat{F}})$ is related to $(\overline{\epsilon_q^2})^{1/2}$ by the formula

$$e_{\hat{F}} = \left\{ \sum_{i=1}^{m} w_i^2 \right\}^{1/2} (\overline{\epsilon_q^2})^{1/2}, \tag{5.9}$$

where w_i is the weight given to the ith observation. $e_{\hat{F}}$ is the rms error on $\hat{F}(q)$, but is more generally known as 'the standard error'. It is obviously quite easy to estimate this standard error quantity once the weights have been calculated, and it is well worth calculating on a once-and-for-all basis the quantity $\left\{ \Sigma_{i=1}^{m} w_i^2 \right\}^{1/2}$ along with the weights themselves (see figure 5.3). It is common practice to present the final calculated estimation of $F(q)$ in the form $\hat{F}(q) \pm e_{\hat{F}}$, which can be interpreted as meaning that there is a roughly 67 per cent chance that the true value of $F(q)$ falls within the range $\hat{F}(q) - e_{\hat{F}}$ to $\hat{F}(q) + e_{\hat{F}}$, these extremes being known as 'standard error limits'.

Equation 5.9 can be used to calculate standard error limits on all the estimated coefficients of the least-squares-plane fit represented by equation 5.3, using the appropriate β weights in place of the w weights in equation 5.9, along with a suitable rms error value for the observed variable q. For kinematic least-squares-plane estimations, such as those given by equations 5.7 and 5.8, the corresponding standard errors may be calculated from

$$e_{\widehat{\text{div}} \mathbf{V}} = e_{\hat{\zeta}} = \left\{ \frac{1}{2} \sum_{i=1}^{m} \left(\beta_{i,1}^2 + \beta_{i,2}^2 \right) \right\} (\overline{\epsilon_{\mathbf{V}}^2})^{1/2} \tag{5.10}$$

where $\overline{\epsilon_{\mathbf{V}}^2}$ is the variance of the vector error on the wind vector $\mathbf{V}$, which though generally considered to be dependent on atmospheric conditions and height as well as the wind-finding system, is reckoned to be typically of the order of 9 m^2 s^{-2} in the lower to middle troposphere.

Let us now consider some standard error limits for least-squares-plane estimations based on arrays of 3 and 7 aerological stations over the British Isles (see figure 5.4). For both arrays, the derivative estimations may be considered as averaged over a length scale of the order of 700 km, which should provide adequate resolution of structures characteristic of synoptic-scale disturbances commonly represented on surface and upper air charts. The sum of the squared weights, and standard error limits corresponding to typical observing errors in mid-tropospheric height and vector winds are shown in table 5.1. These results demonstrate one advantage in using more than the minimum number of observations: *the effect of random observing error decreases as the number of observations increases.* Nevertheless, the errors corresponding to the 7 station array are still

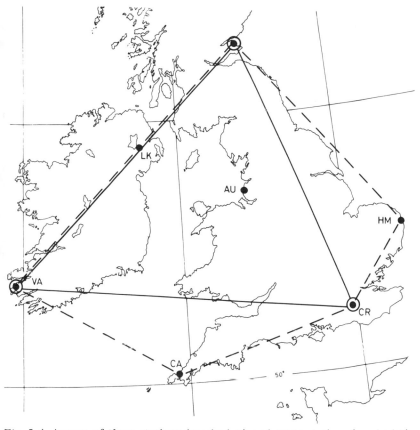

Fig. 5.4 Arrays of three stations (marked ○) and seven stations (marked ●) used to estimate the standard error limits on least-squares-plane estimations associated with random observing errors in upper air observations (see Table 5.1).

Table 5.1 *Weights and error limits in mid-tropospheric height and vector winds.*

Array Number	No. of Stations	Square root of sum of squared weights			rms Errors						
		$\left\{\dfrac{1}{m}\right\}^{1/2}$	$\left\{\Sigma\,\beta^2_{i,1}\right\}^{1/2}$ (m^{-1})	$\left\{\Sigma\,\beta^2_{i,2}\right\}^{1/2}$ (m^{-1})	h' (gpm)	$\mathbf{V}'$ (ms^{-1})	$\hat{h}$ (gpm)	$\hat{\mathbf{V}}$ (ms^{-1})	$\partial h/\partial y$ $(gpm\,m^{-1})$	$\partial\hat{h}/\partial y$ $(gpm\,m^{-1})$	$\widehat{\mathrm{div}\,\mathbf{V}}\,\&\,\hat{\zeta}$ (s^{-1})
1	3	0.58	2.0×10^{-6}	2.3×10^{-6}	12	3	6.9	1.7	2.4×10^{-5}	2.7×10^{-5}	6.4×10^{-6}
2	7	0.38	1.5×10^{-6}	1.7×10^{-6}	12	3	4.5	1.1	1.8×10^{-5}	2.1×10^{-5}	4.9×10^{-6}

quite large. Consider, for example, the standard error in geostrophic wind estimated from equation 5.2 using the least-squares-plane values of $\partial h/\partial x$ and $\partial h/\partial y$. At 500 mb, the corresponding vector standard error $e_{\mathbf{v}g}(= (e_{ug}^2 + e_{vg}^2)^{1/2})$ is 2.3 m s^{-1} for an rms error in h' of 12 geopotential metres. If the estimated geostrophic wind is considered as the average value over the area of analysis, then it is relevant to compare this error with the magnitude of the standard vector error of the mean wind $\bar{\mathbf{V}}$ as estimated from the wind observations themselves. This turns out to be only 1.1 m s^{-1} for an rms vector wind error of 3 m s^{-1}. Thus, it would appear that the error in vector wind estimated by the geostrophic method is considerably larger than the error in the mean vector wind estimation based directly on PILOT wind observations. The reason is not primarily that height data are somehow 'less accurate' than wind data, but rather that the action of *differentiating* the observations with respect to distance effectively 'amplifies' the effect of the observing error. This property of the differencing operation might reasonably be termed the 'curse of meteorological analysis', since many important dynamical quantities can only be estimated after several successive differencing operations, each differencing resulting in a further 'amplification' of what might have been originally considered a rather small error in the data.

This error problem is especially serious in relation to the estimation of horizontal wind divergence div V. In the case of the seven-station array, the standard error on $\widehat{\mathrm{div}}\ \mathbf{V}$ associated with observing errors is about 5×10^{-6} s^{-1}. This turns out to be not much smaller than the root-mean-square variation of div V itself, so that divergence estimations based on the routine PILOT wind data must always be considered very unreliable. (The implications of this result will be discussed in Chapter 6.)

Sampling error

The least-squares-plane analysis represents an attempt to determine the linear part of the x and y components of space variation in the analysed quantity, q. If the variations of q really were perfectly linear along x and y within the analysis area, then the accuracy of $\partial q/\partial x$ and $\partial q/\partial y$ estimations would be limited only by random observing errors. In practice, however, small-scale fluctuations in q, such as might be produced by fronts, thunderstorm circulations, turbulence, etc., generally result in an additional error component. To understand this effect, consider the simplified representation of $q(x)$ in figure 5.5. At any point X_i the difference between q and the value corresponding to the linear trend in q is represented by the residual quantity r_i. Provided that the q surface is sampled at intervals along x, Δx, which are large compared with the scale of the small-scale fluctuations in q, then r_i can be considered as a random quantity with zero mean value when averaged over the whole region. Thus, if q is observed at a very large number of points, these 'random fluctuations' about the true linear trend tend to cancel out in the estimation of $\bar{q}$ and $\partial q/\partial x$. However, if the q surface is sampled at only a few points, then these residuals may result in quite large errors of estimation. In a sense, the sampling error effect is therefore similar to the random observing error effect, except that, in the case of sampling error, the

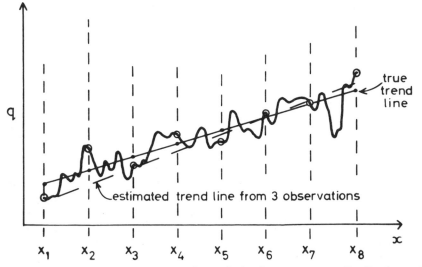

Fig. 5.5 A schematic representation of the instantaneous distribution of an observable q with respect to distance x as sampled by observing stations at $X_1, X_2 \ldots X_8$. The linear trend in q as estimated from observations (○) at X_1, X_5 and X_8 is clearly in error compared with true linear trend, whereas if all 8 observations were used the estimated trend line would be close to the true trend line.

'errors' are not associated with the observing system, but rather with the nature of the atmosphere itself.

The standard error limits on the estimated quantity $\widehat{F}(q)$ in equation 5.1 resulting from the combined effect of random observing error *and* sampling error can be estimated from the calculated differences between fitted and observed values of q at all m points used in the analysis. If such an estimation is made, then it is not actually necessary to know the value of the observing error variance. Unfortunately, such an analysis is more involved than that described under random observing error on p. 62. Moreover, the probability that the true value of $F(q)$ falls within the range $\bar{F}(q) \pm e_{\widehat{F}}$ is in this case a function of the number of observations, so that standard error limits in themselves do not give an adequate indication of the accuracy of estimation. Incidentally, no estimation of this type is possible if the number of observations is equal to the number of unknown coefficients or weights in the analysis model. In that case, standard error limits can only be estimated in terms of an assumed value of observing error variance.

For these reasons, it is much simpler in practice to stick to an error analysis based on observing error effects alone, and simply accept that the corresponding standard error limits are probably somewhat optimistic. Before leaving this problem it is, however, important to note that both observing and sampling error effects decrease as the number of observations used (m) increases. Thus it is desirable that the number of observations used (m) should well exceed the number of weights or coefficients (n), a useful 'rule-of-thumb' being 'of the order of or greater than $2 \times n$'. This criterion may often not be easy to satisfy when analysing upper air data in 'data-sparse' regions, since the corresponding area of analysis may thus be comparable with, or larger than, the scale of atmospheric structure being investigated. In this case, the third type of error becomes important.

Analysis model error

This error is illustrated schematically in figure 5.6. It results from using the wrong 'model' for the estimated shape of the surface being analysed. Thus, in figure 5.6, a straight-line fit to four observations scattered over a large region of map space results in a quite unrealistic estimated $q(x)$ line, and a very large error in $\partial q/\partial x$ at the centre of the array x_0. In the case of the least-square-plane fit, this type of error is probably not as important as the effect of random observing and sampling errors, provided that the analysis area is not too large, e.g. not larger than the order of 900×900 km in the case of synoptic-scale analysis. Clearly, the effect

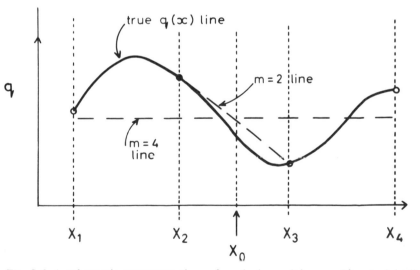

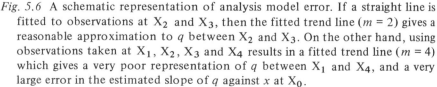

Fig. 5.6 A schematic representation of analysis model error. If a straight line is fitted to observations at X_2 and X_3, then the fitted trend line ($m = 2$) gives a reasonable approximation to q between X_2 and X_3. On the other hand, using observations taken at X_1, X_2, X_3 and X_4 results in a fitted trend line ($m = 4$) which gives a very poor representation of q between X_1 and X_4, and a very large error in the estimated slope of q against x at X_0.

of interpolation error can be reduced by choosing a more appropriate model to describe the spatial variation of the quantity being analysed, and this possibility is discussed in Chapter 7.

Summary

The basic requirements of map analysis in relation to dynamical and kinematic estimations have been described, together with a simple objective analysis scheme which can be effected using a pocket calculator. Unfortunately, the errors of estimation associated with this type of analysis are often of similar magnitude as the variable being estimated, in which case, great care must be taken when interpreting results. Nevertheless, I shall demonstrate in the next chapter that meaningful estimations of dynamical and kinematic processes can be obtained from such a procedure.

References

Bellamy, J. C. (1949) 'Objective calculations of divergence, vertical velocity, and vorticity', *Bull. Am. Met. Soc.*, 30, 45–9.

Dahlquist, G. and Björck, A. (1974) *Numerical Methods*, Chapter 4, Englewood Cliffs, N. J., Prentice-Hall Inc.

Gadd, A. J. (1981) 'Numerical modelling of the atmosphere', this volume, 194–204.

Harwood, R. S. (1981) 'Atmospheric vorticity and divergence', this volume, 33–54.

Panofsky, H. A. (1981) 'Atmospheric hydrodynamics', this volume, 8–20.

Saucier, W. J. (1955) *Principles of Meteorological Analysis*, Chicago, University of Chicago Press, 325–6.

6
Practical analysis of dynamical and kinematic structure: some applications and a case study

M. A. PEDDER
University of Reading

In Chapter 5, I described the least-squares-plane analysis method for estimating the spatial gradients of an observed meteorological variable with respect to horizontal x(east) and y(north) co-ordinate directions. This objective analysis procedure is sufficiently simple to be effected with the aid of an ordinary pocket calculator using surface or aerological data drawn from a small number of fixed reporting stations. In this chapter, I first discuss some typical applications of these measurements which are relevant to a quantitative understanding of some of the dynamical and kinematic processes described in earlier chapters, and then illustrate some of the methods involved by presenting a case study of anticyclonic development over the British Isles.

Some applications of least-squares-plane estimations

Although the least-squares-plane analysis is limited to the estimation of local first-order space derivatives of height, pressure, vector wind, etc., it is still possible to carry out some interesting dynamical and kinematic measurements based on this technique.

The geostrophic wind

A simple, but by no means trivial measurement is the estimation of the geostrophic wind vector V_g which is related to the horizontal gradient of geopotential height or pressure, as described in Chapter 5 and also by Panofsky (1981a). Routine comparison between this estimate and the mean wind $\bar{V}$ given by PILOT data could be used to test the validity of the geostrophic approximation for the area of interest. Alternatively, estimates of V_g at some standard analysis level, e.g. 500 mb, could be compared with a simple grid point differencing estimation (see Gadd, 1981), where grid point values have been interpolated 'by eye' off a published field analysis, such as found in the *European Weather Bulletin*.

Kinematic measurements

At a rather more ambitious level, vertical profiles of vorticity, divergence, vertical motion (see below), and the important divergence term in the vorticity equation (see Harwood, 1981) can all be estimated on a time series basis and related to sequences of surface observations of local weather characteristics.

The estimation of synoptic-scale vertical motion is particularly exciting, since this variable not only provides, with $\bar{u}$ and $\bar{v}$, a three-dimensional view of the local motion field, but is also most directly associated with the 'visible' weather elements of mid-tropospheric cloud cover and widespread precipitation. As described by Panofsky (1981a), vertical motion is related to horizontal wind divergence div $\mathbf{V}$ by the continuity equation, and a convenient form of this equation for practical estimations is

$$\omega_p = \omega_0 + \int_{p_0}^{p} \text{div } \mathbf{V} \, dp, \quad p < p_0, \tag{6.1}$$

where div $\mathbf{V}$ is here defined on a constant pressure surface and $\omega \ (= dp/dt)$ is the pressure change following the motion, which is related approximately to vertical velocity $w = dz/dt$ by

$$\omega = -\rho g w, \tag{6.2}$$

where ρ is the density of the air at the level p. In equation 6.1 the subscript $_0$ is used to denote a mean-sea-level pressure surface value. In practice, ω_0 is normally assumed to be zero, since total surface pressure tendency is invariably much smaller than ω typical of the middle troposphere. If div $\mathbf{V}$ has been estimated on a number of standard pressure levels, then ω_p can be estimated most simply by trapezoid-rule integration, which follows the rule

$$\hat{\omega}_j = \hat{\omega}_{j-1} + \frac{\Delta p}{2} (\widehat{\text{div }} \mathbf{V}_{j-1} + \widehat{\text{div }} \mathbf{V}_j), \tag{6.3}$$

where $(j-1)$ and j identify pressure surfaces p_{j-1} and $p_j (p_{j-1} > p_j)$, $\Delta p_j = p_{j-1} - p_j$, and $\hat{\omega}_0 = 0$. Unfortunately, the relatively large errors typical of div $\mathbf{V}$ often then result in quite unrealistic profiles of ω_j. To see why this might be so, consider five values of div $\mathbf{V}$ at equal pressure intervals between p_0 and 500 mb, which each have a standard error of 5×10^{-6} s^{-1}. The corresponding standard error in $\hat{\omega}$ at 500 mb as estimated from equation 6.3 is about 1.2×10^{-3} μb s^{-1} (nearly 2 cm s^{-1}), which is not much less than the synoptic-scale averaged magnitude of vertical motion typical of mid-latitude weather systems*. More-over, the absolute error in $\hat{\omega}$ tends to increase with pressure height, so that the estimated profile often does not tend towards zero vertical motion in the upper atmosphere, as illustrated in figure 6.1a.

* ω may also be estimated from other indirect methods which do not involve the integration of kinematic divergence estimates. All these methods are susceptible to relatively large errors of estimation, but are generally consistent with respect to the resulting order of magnitude and variability of synoptic-scale averaged vertical motion.

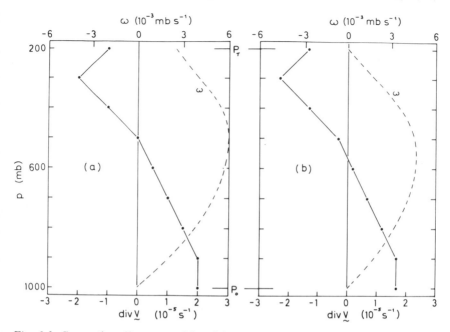

Fig. 6.1 Correcting divergence bias. (a) The type of vertical motion profile ($\hat{\omega}$) obtained by integrating the estimated divergence profile ($\widehat{\text{div}}\ V$) assuming only that $\hat{\omega} = 0$ at p_0. (b) The profiles obtained after adjusting the estimated divergence such that $\hat{\omega} = 0$ at p_0 and p_T.

More realistic $\hat{\omega}$ profiles can be obtained by applying a 'bias correction', a technique which nicely illustrates how synoptic estimations can be improved by appealing to other well-known physical or dynamical characteristics of the atmosphere. In this case, the characteristic of interest is that immediately above tropopause level any significant vertical motion tends to produce very rapid heating or cooling of the air at that level, this being a consequence of adiabatic compression or expansion in a near-isothermal layer (see Panofsky, 1981b). Observations suggest that such rapid changes of temperature do not normally take place at this level, so that we might reasonably conclude that ω is normally very small just above the tropopause. This implies that if p_T is the pressure just above the tropopause, then

$$\omega_T = \int_{p_0}^{p_T} \text{div } V\ dp \simeq 0. \tag{6.4}$$

We can use this result to correct our original $\hat{\omega}$ profile as follows: Supposing $\hat{\omega}_T$ is the value of ω at p_T as found from equation 6.3, then the quantity $\hat{\omega}_T/(p_0 - p_T)$ represents the mean tropospheric divergence between p_0 and p_T. If we

Dynamical Meteorology

subtract this quantity from all the original values of div **V** and integrate equation 6.3 a second time, then the new profile of $\hat{\omega}$ goes to zero at p_T, as illustrated in in figure 6.1b. This procedure is frequently termed the 'divergence bias correction'; it does not guarantee that the estimated profile of ω is correct, but it does give more realistic profiles near p_0 and p_T as well as reducing the probable error in the middle troposphere, provided of course that the condition '$\omega = 0$ at p_T' is more or less valid.

The above bias correction scheme is 'optimum' if the standard error limits on $\widehat{\text{div}}$ **V** are the same at all levels. O'Brien (1970) has described a more elaborate bias correction scheme which attempts to take into account the increase in wind observing error with height. However, it should be remembered that sampling and analysis model errors may also be important in practice, and these do not necessarily either increase or decrease with height. Moreover, sampling and model errors at different levels need not be independent, since the overall wind field structure may not vary much with height, even though its divergent component varies considerably. For these reasons, there is probably not much point in applying numerically involved bias correction schemes when divergence estimations are based on relatively few station observations.

This bias correction is not only useful in relation to the estimation of ω itself, but also for providing more reliable mean layer divergences when this quantity is involved in the measurement of other meteorological processes such as the divergence term in the vorticity equation $(-\text{div }\mathbf{V}(f + \zeta))$, and water vapour convergence $(-\bar{Q}\,\text{div }\mathbf{V}$, where $\bar{Q}$ is area-averaged specific humidity). In such estimations, a 'corrected' layer mean divergence is obtained as $\overline{\text{div }\mathbf{V}} = (\hat{\omega}_2 - \hat{\omega}_1)/(p_1 - p_2)$, where $\hat{\omega}_2$ and $\hat{\omega}_1$ are the bias-corrected ω estimations at levels p_2 and p_1.

The thermal wind relation

Apart from these purely kinematic estimations, the least-squares-plane analysis can also be used to investigate horizontal temperature gradients and the rate of horizontal thermal advection, the latter being an important feature of baroclinic disturbances in middle latitudes. Temperature gradients at standard reporting levels could of course be estimated directly from reported temperatures, but more reliable layer-averaged estimations may be obtained from the standard level geopotential height data.

From the hydrostatic equation (see Panofsky, 1981a), the mean temperature, $\bar{T}$, of the atmosphere between two pressure surfaces p_1 and p_2 $(p_1 > p_2)$ is proportional to the geopotential *thickness* of the layer $\delta h = (h_2 - h_1)$, this relationship being represented by

$$\bar{T} = \frac{g\,\delta h}{R\,\ln(p_1/p_2)}, \qquad (6.5)$$

where R is the specific gas constant for air (about 287 J kg^{-1} K^{-1}) and $g = 9.80665$ m s^{-2}. The magnitude of layer mean horizontal thermal advection, calculated as $(\bar{u}\,\partial\bar{T}/\partial x + \bar{v}\,\partial\bar{T}/\partial y)$ could thus be estimated from geopotential

72

height data alone if $\bar{u}$ and $\bar{v}$ are assumed to be the same as their geostrophic values.

In the middle and higher latitudes, the horizontal components of temperature gradient can also be estimated from wind data alone, since in geostrophic balance the vertical shear of the horizontal wind is uniquely related to horizontal gradients of temperature (see Panofsky, 1981a). Thus, if u_1, v_1 and u_2, v_2 are the components of wind observed at pressure levels p_1 and p_2 ($p_1 > p_2$) then the horizontal components of layer mean temperature gradient can be estimated from

$$\frac{\partial \bar{T}}{\partial x} = \frac{f(v_2 - v_1)}{R \ln(p_1/p_2)} \tag{6.6a}$$

and

$$\frac{\partial \bar{T}}{\partial y} = \frac{f(u_1 - u_2)}{R \ln(p_1/p_2)} \tag{6.6b}$$

these equations representing one form of the 'thermal wind relation'. Using area-averaged values of u and v in place of u_1, v_1, u_2, v_2 in equations 6.6a and 6.6b, it is thus possible to compare the rate of horizontal thermal advection based on the geostrophic assumption with that obtained more directly from the wind data and station thicknesses alone. Such a comparison is by no means trivial, since the assumption that the actual rate of horizontal thermal advection is virtually the same as that produced by the geostrophic wind forms the basis both for interpreting 1000/500 mb height—thickness charts and for inferring the horizontal distribution of mid-troposphere vertical motion in middle latitudes from the instantaneous height and thickness distribution.

Reducing the errors

In all the above examples, the reliability of each individual estimation is limited by the various errors of estimation described in Chapter 5. In practice, the effect of these errors may be so large that often little confidence can be placed on any single measurement which involves a differential estimation. There are, however, two ways in which the effect of these errors can be reduced. One of these has already been described in relation to the estimation of vertical motion by the kinematic method, namely the addition of the 'constraint' that the estimated ω profile should be zero at the surface and near the tropopause level. The scope for applying constraints of this type to simple least-squares-plane estimations is very limited, but this principle is important in relation to more sophisticated analysis schemes, and is discussed briefly in Chapter 7.

The second method for reducing the effect of errors is to apply some kind of averaging, either to the observations themselves or to the estimations. Using the geopotential thickness of a layer in place of temperature at a single level is an example of averaging in the *vertical*; (the reported geopotential heights are themselves obtained by integrating the detailed profile of temperature recorded by the radiosonde with respect to pressure, a process which is equivalent to averaging the observed temperatures.) Averaging in the vertical can be an effective method of reducing the influence of random observing and sampling errors, though obviously at the expense of detail in the vertical structure represented by the final analysis. If standard pressure level data are used, an appropriate averaging procedure is to calculate the pressure weighted mean. Thus suppose we have three measurements $\hat{q}_1$, $\hat{q}_2$, $\hat{q}_3$ corresponding to the 850, 700 and 500 mb levels, then the pressure weighted mean for the layer 850–500 mb is calculated as

$$\hat{\bar{q}}^p = \frac{75\hat{q}_1 + (75 + 100)\hat{q}_2 + 100\hat{q}_3}{(850 - 500)}, \tag{6.7}$$

the multiplying factors being the pressure intervals defined by dividing each layer into two equal parts and calculating the total interval 'surrounding' each quantity. This can obviously be extended to include many levels, and we again note that this type of estimation might also be written

$$\hat{\bar{q}}^p = \sum_{i=1}^{m} w_i\hat{q}_i, \tag{6.8}$$

which is similar in form to equation 5.1. If $e_{\hat{q}}$ is the estimated standard error in $\hat{q}$, it then follows that the standard error in $\hat{\bar{q}}$ is given by

$$e_{\hat{\bar{q}}} = e_{\hat{q}} \left\{ \sum_{i=1}^{m} w_i^2 \right\}^{1/2}. \tag{6.9}$$

In practice, the effect of averaging estimations on the standard pressure levels over 500 mb is to reduce the random error component by a factor of about times 2.

Estimations may also be averaged in time to provide a 'smoothed' or 'filtered' time series representation in which the effect of random fluctuations may be reduced considerably. Little detail of the time evolution of synoptic-scale processes is lost by averaging measurements over one day. For 12-hourly data the corresponding three-point moving average centred on t_0 is

$$\hat{\bar{q}}_0 = \frac{1}{4}\hat{q}_{-12} + \frac{1}{2}\hat{q}_0 + \frac{1}{4}\hat{q}_{+12}, \tag{6.10}$$

which reduces the effect of random errors of a factor of times 1.6. For 6-hourly data a five-point moving average is given by

$$\hat{\bar{q}}_0 = \frac{1}{8}\hat{q}_{-12} + \frac{1}{4}\hat{q}_{-6} + \frac{1}{4}\hat{q}_0 + \frac{1}{4}\hat{q}_{+6} + \frac{1}{8}\hat{q}_{+12}. \tag{6.11}$$

This reduces the effect of random errors by a factor of times 2.1. (Rather better time-filtering techniques suitable for meteorological analysis are described by Craddock, 1968.)

Finally, the data or results obtained during several periods of rather similar synoptic structure can be averaged, using an ordinary arithmetic mean, to form a 'composite' structure. If m values of $\hat{q}$ are used to estimate the composite value $\hat{\bar{q}}$, then the effect of random error in $\hat{q}$ is reduced by a factor of $m^{1/2}$. This technique has been applied with various degrees of sophistication in many important research studies. In practice it is not easy to decide what constitutes a 'similar situation', and if attention is restricted to the area covered by a small number of aerological stations a very large number of periods may have to be analysed before a few similar cases of the type required are actually found.

Case study of anticyclonic development over the British Isles

In the following sample case study, several of the above-mentioned techniques are used to investigate the kinematic and dynamical structure of the atmosphere up to 100 mb over the British Isles during late February 1979. Space does not allow an analysis based on all the possible least-squares-plane estimations, and only the results based on the analysis of PILOT wind data are presented here.

Synoptic situation

Between 21 and 28 February 1979, a mid-Atlantic surface anticyclone extended towards the British Isles and developed to produce record high msl pressures (1045 mb) over central England during the 25th. This surface development appears to have taken place initially beneath the right exit of a strong west to north-westerly upper tropospheric jet. By 24 February the upper jet had developed a characteristic 'split', with one branch curving anticyclonically over the North Sea and flowing southwards across the Iberian peninsula, while the second branch curved cyclonically over Scandinavia. After 0000 GMT on the 25th, the upper ridge in the height field gradually relaxed and tilted more south-west to north-east, as a developing upper trough propagated rapidly across the central North Atlantic. The upper ridge rapidly collapsed during the 26th and 27th and by 28 February was replaced over the British Isles by the trough pattern. The corresponding sequence of 1200 GMT surface charts, reproduced from the Weather Log published by *Weather,* is shown in figure 6.2.

The purpose of the following study was to obtain kinematic time series representing tropospheric changes in the circulation over the British Isles, and to attempt to isolate the main dynamical mechanism responsible for the development of the intense anticyclonic circulation in this region.

The analysis

A kinematic least-squares-plane analysis was applied to PILOT wind data taken from the station array shown in figure 6.3. These data were available every 6 hours and, where possible, were analysed at every standard pressure level up to 100 mb. The surface and 900 m winds were taken as corresponding to the nearest equivalent pressure surface at Aughton (near the centre of the array). In estimating vertical motion, the constraints $\omega = 0$ at the msl pressure and $\omega = 0$ at

Fig. 6.2 Mid-day (GMT) msl analysis charts between 21 and 28 February 1979, showing the development and eventual decay of an anticyclone over the British Isles.

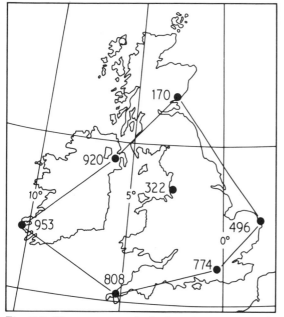

Fig. 6.3 The seven aerological station array used
for the least-squares-plane analysis of standard
level PILOT data.

$p = 200$ mb were assumed throughout the period of study, the bias correction
described by Pedder (1981) being applied, assuming that the error in div $\mathbf{V}$
estimations did not vary with height. Standard error limits on all estimated
quantities and processes were calculated assuming an rms random vector wind
observing error of 3 m s^{-1} throughout the troposphere, and neglecting the effect
of sampling and analysis model errors*.

Results

Figure 6.4 shows four sample profiles of $\hat{\zeta}$, $\widehat{\text{div }\mathbf{V}}$ (before bias correction) and $\hat{\omega}$
against pressure height at 1200 GMT between 22 and 25 February. In all four
cases, it is worth noting that the standard error limits on $\hat{\zeta}$ are small compared
with the typical magnitude of $\hat{\zeta}$, so that we can be fairly confident that these
profiles are representative of the true area-averaged relative vorticity profiles.
The same cannot be said of the divergence profiles, for which the magnitude of

---

* It is perhaps fair to mention that all these estimations were carried out on a mini-computer
rather than a pocket calculator. I estimate that the total time involved was of the order of
10 hours, about half of which was spent decoding, copying out and checking the PILOT
wind data.

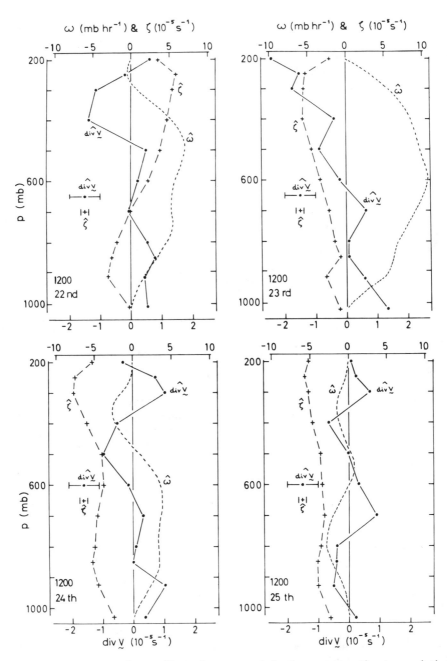

Fig. 6.4 Four sample profiles of uncorrected divergence estimates, relative vorticity and bias-corrected vertical motion as $\omega = dp/dt$. Error bars represent standard error limits calculated for a random wind observing error of 3 m s^{-1}.

$\widehat{\text{div } \mathbf{V}}$ is rarely much greater than its standard error. Nevertheless, the bias-corrected profiles of $\bar{\omega}$ are generally consistent with that which might be expected during the development of the anticyclone (the positive mid-troposphere ω values corresponding to subsidence).

The estimated vorticity profiles plotted against time were used to form a time–height cross-section of vorticity changes over the British Isles between 22 and 28 February, as shown in figure 6.6, which can be compared with a similar cross-section of the horizontal wind vectors observed over Aughton, as shown in figure 6.5. From these sections, we note how a rapid increase of anticyclonic vorticity throughout the troposphere took place between 1200 GMT on the 22nd and 0000 GMT on the 24th, corresponding to a period of maximum upper wind speeds in the north-westerly flow over central England. The increase in anticyclonic vorticity in the lower troposphere is seen to be weaker than in the upper troposphere, with a definite time lag or 'tilt' being present in the maximum values, as indicated by the dashed line in figure 6.6. Relatively little change in $\hat{\zeta}$ below 300 mb took place between 1200 GMT, 24th, and 1200 GMT, 27th, but the arrival of the upper trough over the British Isles by the 28th is indicated by the rapid change towards positive vorticity after 0000 GMT on the 27th.

Superimposed on the vorticity time–height section are arrows representing the direction and magnitude of the estimated vertical motion at 500 mb. Despite the relatively large possible error in this quantity, the overall pattern is consistent with the dynamical structure typical of such mid-latitude disturbances. In partic-

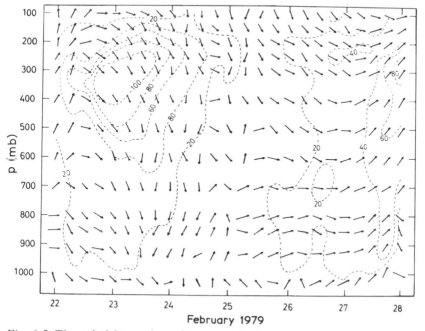

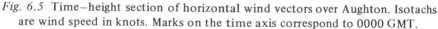

Fig. 6.5 Time–height section of horizontal wind vectors over Aughton. Isotachs are wind speed in knots. Marks on the time axis correspond to 0000 GMT.

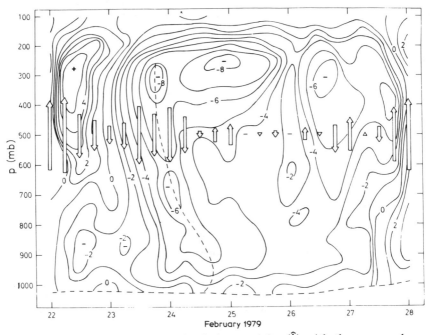

Fig. 6.6 Time–height section of relative vorticity ($\hat{\zeta}$) with the sense and magnitude of mid-tropospheric vertical motion represented by large arrows. Isopleths of $\hat{\zeta}$ are in units of 10^{-5} s^{-1}. The magnitude of vertical velocity at 500 mb is proportional to the length of an arrow; the longest (0600 GMT on 22nd) corresponds to about 18 cm s^{-1}.

ular, we notice that the strongest vertical motion takes place during periods when the circulation, as represented by vorticity, is undergoing *change*, rather than when it is at its maximum or minimum. Strong vertical motion corresponds to appreciable ageostrophic accelerations in the local motion field, these being mainly responsible for the horizontal wind divergence, as explained by Harwood (1981). Hence, the observed local changes might, in part, have been associated with ageostrophic 'development' processes. This possibility can be explored further by substituting the kinematic estimations of divergence and vorticity into the vorticity equation as presented by Harwood (1981) and discussed further by Green (1981). From these discussions we might reasonably suppose that the principal mechanisms affecting the local synoptic-scale changes in the vertical component of relative vorticity ($\partial\zeta/\partial t$) are (1) the horizontal advection of absolute vorticity, and (2) the changes in absolute vorticity following the motion associated with horizontal divergence (or 'vertical shrinking'), represented by the term $-$ div $\mathbf{V}$ $(f + \zeta)$. Now we cannot estimate effect 1 directly from the least-squares-plane estimations, since the horizontal gradient of relative vorticity requires measurement of higher-order wind derivatives such as $\partial^2 u/\partial x\partial y$. However, the advection process can be estimated as a *residual* between the observed local rate of change $\partial \hat{\zeta}/\partial t$ and the term $-\widehat{\text{div } \mathbf{V}} \,(f + \hat{\zeta})$ from the vorticity equation written in the form

$$\frac{\partial \zeta}{\partial t} \simeq [\text{absolute vorticity advection}] - \text{div } \mathbf{V}\{f + \zeta\}. \qquad (6.12)$$

We could then compare the magnitude of the latter term with the estimated advection process, and thus come to some conclusion as to the relative importance of 'advection' versus 'baroclinic development' processes.

At this point, however, we are again up against the error problem. Because the standard error in $\widehat{\text{div } \mathbf{V}}$ is comparable with the magnitude of variation in $\text{div } \mathbf{V}$ itself, the errors in $-\widehat{\text{div } \mathbf{V}} \,(f + \hat{\zeta})$ also tend to be of the same order as this term. Considerable errors of interpretation could thus result if measurements were based directly on the raw divergence and vorticity estimations. More realistic results can be obtained by averaging in the vertical and in time. In this study, $\widehat{\text{div } \mathbf{V}}$ was therefore replaced by a layer mean based on the bias-corrected $\hat{\omega}$ profile, and ζ by its pressure-weighted layer mean. The term $-\widehat{\text{div } \mathbf{V}} \,(f + \hat{\zeta})$ was then averaged over 24 hours using the 5-point moving average filter, and compared with the corresponding 24-hour difference in pressure-weighted mean vorticity. The layers selected were 900–600 mb, representing processes in the lower troposphere above the frictional boundary layer, and 500–200 mb, representing processes in the upper half of the troposphere. The results are shown in the form of time series in figures 6.7 and 6.8.

Figures 6.7a and 6.8a show the smoothed time series of divergence $(\widehat{\text{div } \mathbf{V}})$ and vorticity $(\hat{\zeta})$ for the two layers. These series again show how the magnitude of the ageostrophic flow, as measured by $\overline{\text{div } \mathbf{V}}$, is associated with *changes* in vorticity, this correlation being most noticeable in the upper troposphere. It is interesting to notice that the maximum anticyclonic vorticity in both layers occurs about 24 hours before the time of maximum surface pressure over the British Isles, but coincides quite well with the maximum surface development of the overall pressure pattern on the 23rd (see figure 6.2).

Figures 6.7b and 6.8b show the smoothed time series of the divergence term $- \text{div } \mathbf{V} \,(\zeta + f)$ compared with 24-hour difference estimations of $\partial \zeta/\partial t$. Where $\partial \zeta/\partial t$ is much more positive than $- \text{div } \mathbf{V} \,(\zeta + f)$, then this suggests significant positive vorticity advection; if $- \text{div } \mathbf{V} \,(\zeta + f)$ is much more positive than $\partial \zeta/\partial t$, this suggests significant negative vorticity advection. Following from this assumption, we notice important differences between the vorticity budgets of the upper and lower trophospheric layers, particularly during the period of strong anticyclonic development 1200 GMT, 22nd, to 1200 GMT, 24th, but also after 0000 GMT, 27th and, to a lesser extent, before 1200 GMT, 22nd. In the upper troposphere, the divergence term is generally opposite in sign to $\partial \zeta/\partial t$, and the difference between the two terms is large when one or both of these terms is large. When $\partial \zeta/\partial t$ is *negative* (vorticity becoming more anticyclonic with time), then the residual suggests an even larger rate of *negative vorticity advection*, this situation being reversed when $\partial \zeta/\partial t$ is *positive*. Thus, in the upper troposphere, most of the local vorticity changes are seen to be associated with the process of

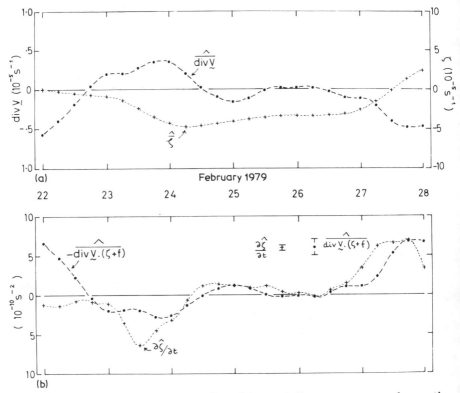

Fig. 6.7(a) Smoothed time series of vorticity and divergence averaged over the layer 900–600 mb between 0000 GMT, 22nd and 0000 GMT, 28 February 1979.

Fig. 6.7(b) Smoothed time series of the vorticity generation term $-\text{div }\mathbf{V}(\zeta + f)$, and the observed local rate of change of vorticity $\partial\zeta/\partial t$ (as estimated from 24-hour differencing) averaged over the layer 900–600 mb.

vorticity advection during periods of strong upper flow, with the divergence term tending to be opposite in sign but much smaller than the vorticity advection term. The maximum rate of vorticity advection during the period of anticyclonic development occurs around 0000 GMT, 23rd. At this time, the layer mean rate of vorticity advection corresponds to a local change in ζ of about -0.6×10^{-5} s^{-1} h^{-1}, and the mean divergence term to about $+0.2 \times 10^{-5}$ s^{-1} h^{-1}, resulting in the observed net change of about -0.4×10^{-5} s^{-1} h^{-1}.

In the lower tropospheric layer the difference between the divergence term and $\partial\zeta/\partial t$ is much smaller than in the upper troposphere. During the period of strong anticyclonic development, $-\overline{\text{div }\mathbf{V}(\zeta + f)}$ and $\overline{\partial\zeta/\partial t}$ tend to be of the same sign, this being also true during the period of cyclonic development after 1800 GMT, 26th. During 23rd, the observed local rate of change in ζ can be attributed to almost equal contributions from *negative vorticity advection* and

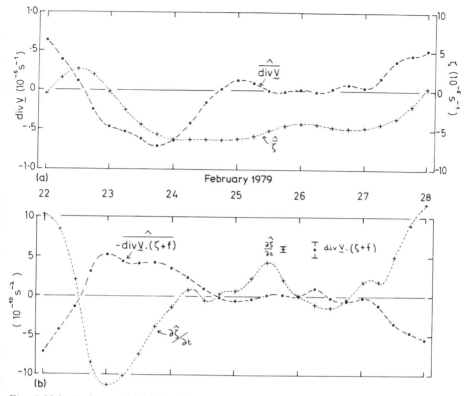

Fig. 6.8(a) As figure 6.7(a) but for the layer 500–200 mb.
Fig. 6.8(b) As figure 6.7(b) but for the layer 500–200 mb.

'*vertical shrinking*', i.e. − div **V** $(\zeta + f)$ negative. Between 1800 GMT, 26th and 1800 GMT, 27th, the observed local increase can be attributed to almost equal contributions from *positive vorticity advection* and '*vertical stretching*', i.e. − div **V** $(\zeta + f)$ positive. Thus, in the lower troposphere, the ageostrophic development process appears to have played an important role in the local development and decay of the anticyclone as seen in the surface pressure pattern.

Conclusions: a 'quasi-geostrophic' interpretation

In the following discussion, I shall concentrate on the anticyclonic development period 1200 GMT, 22nd to 0000 GMT, 24th, and I shall attempt to demonstrate how the relatively simple kinematic measurements can be used to bring together many of the dynamical, kinematic and thermodynamic concepts presented in earlier chapters in order to arrive at a more complete diagnostic interpretation of the results.

In middle latitudes, the tendency towards geostrophic balance results in 'ridging' or 'upward bulging' of pressure surfaces where there is negative relative vorticity, and 'troughing' or 'downward bulging' of pressure surfaces where there is positive relative vorticity. From the two-layer time series results (figures 6.7 and 6.8) we have concluded that upper level negative vorticity advection was locally much stronger than the lower level advection. Thus, if we imagine an observer moving along with the mid-tropospheric flow during this period, we might have expected him to see the changes in shape of the upper and lower pressure surfaces represented in figure 6.9, due to the vertical difference in the rate of horizontal vorticity advection (or 'differential vorticity advection'). A consequence of this change is seen to be an increase in the *thickness* of the atmosphere in the vicinity of the observer relative to that on either side. It follows from the hydrostatic balance approximation that the temperature of the air in the vicinity of the observer must have increased relative to the air on either side during this time. One way in which such a temperature change could take place is by adiabatic heating during subsidence of this central air column (see Panofsky, 1981b). This picture is actually consistent with the $\hat{\omega}$ profiles found at this time over the British Isles. This link-up between the observed vorticity changes and the observed vertical motion is part of the so-called 'quasi-geostrophic theory' used to understand dynamical structures in middle latitudes. The term 'quasi' is used here because we are associating a change in height−field distribution as required to *maintain geostrophic balance* with an *ageostrophic process*, namely the horizontal wind divergence required to produce the vertical motion which is in turn required to satisfy the first law of thermodynamics ('heat produced is equivalent to work done'). If we think of the observed vertical motion and divergence as a response to a 'forcing' provided by the differential vorticity advection, then the development of low level anticyclonic vorticity and the surface anticyclone may also be thought of as a response to this forcing

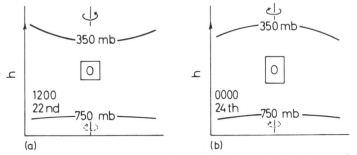

(a)   (b)

Fig. 6.9 Schematic representation of the change in the shape of pressure surfaces above and below an observer (O) moving with the mid-tropospheric flow, corresponding to differential vorticity advection in a geostrophically balanced flow. Above him, negative vorticity advection is large, so that he sees the upper troposphere becoming more anticyclonic with a corresponding upward bulge in the *h* surface. This effect is small below him. Note the subsequent change in the distribution of 350−750 mb thickness in the vicinity of the observer.

mechanism, since the conservation of potential vorticity principle (see Harwood, 1981) requires that $(\zeta + f)$ must decrease where there is horizontal wind divergence. (Incidentally this $-$ div $\mathbf{V}\,(\zeta + f)$ effect can also be seen partially to offset the tendency for the vertical gradient of negative vorticity to increase in time in response to differential vorticity advection.)

Although the above interpretation is consistent with the observations, it is not in itself a sufficient description of the dynamical characteristics of such a development. Referring again to figure 6.9 we could also 'explain' the expected change in thickness distribution by having a horizontal distribution of layer mean temperature such that the rate of positive thermal advection by the horizontal wind in the vicinity of the observer was greater than that on either side, i.e. his column of air could be moving into a region (figure 6.9b) where the air to either side was relatively cooler than in figure 6.9a. In such a situation, there would be no need for adiabatic heating, and therefore no need for subsidence in the vicinity of the observer. The reader can probably guess that if there were no differential vorticity advection, then such a pattern of thermal advection could actually result in ascending motion in the vicinity of the observer — though I leave it as an exercise for the enterprising reader to discover how 'quasi-geostrophic' theory 'explains' this result.

More sophisticated treatments of quasi-geostrophic theory, such as those presented by Holton (1972) and Hoskins *et al.* (1978) suggest that, most of the time, the effects of differential vorticity advection along the vertical and variations in thermal advection in the horizontal tend to be in opposition, so that the sign of the observed vertical motion is determined by the residual between these opposing forcing mechanisms. In the case of the anticyclonic development studied here, the sign of the estimated vertical motion is consistent with the sign of the estimated differential vorticity advection, and we might reasonably conclude that differential vorticity advection was the dominating 'quasi-geostrophic forcing mechanism' responsible for the observed increase in low level anticyclonic vorticity.

Comments

The above theoretical arguments are admittedly difficult to follow without recourse to a more rigorous mathematical representation of the processes involved. Nevertheless, I hope that the reader may at least be able to appreciate how relatively simple measurements of dynamical and kinematic processes can lead to a much deeper understanding of 'how the atmosphere works' than would ever be possible merely by studying surface and upper-level analysis charts.

There are, of course, some important deficiencies in the least-squares-plane estimations used for the above case study, notably the inability actually to measure differential vorticity advection or horizontal variations in thermal advection. Such processes, along with many other important dynamical quantities, require the estimation of higher-order space derivatives than those available from the least-squares-plane analysis. Some analysis techniques which can provide such estimations are described in the next chapter.

References

Craddock, J. M. (1968) *Statistics in the Computer Age*, London, English Universities Press,Chapter 15.

Gadd, A. J. (1981) 'Numerical modelling of the atmosphere', this volume, 194–204.

Green, J. S. A. (1981) 'Trough–ridge systems as slant-wise convection', this volume, 176–193.

Harwood, R. S. (1981) 'Atmospheric vorticity and divergence', this volume, 33–54.

Holton, J. R. (1972) *An Introduction to Dynamic Meteorology*, London, Academic Press, Chapter 7.

Hoskins, B. J., Draghici, I. and Davies, H. C. (1978) 'A new look at the ω-equation', *Quart. J. R. Met. Soc.*, 104, 31–8.

O'Brien, J. J. (1970) 'Alternative solutions to the classical vertical velocity problem', *J. appl. Met.*, 9, 197–203.

Panofsky, H. A. (1981a) 'Atmospheric hydrodynamics', this volume, 8–20.

Panofsky, H. A. (1981b) 'Atmospheric thermodynamics', this volume, 21–32.

Pedder, M. A. (1981) 'Practical analysis of dynamical and kinematic structure: principles, practice and errors', this volume, 55–68.

7

Practical analysis of dynamical and kinematic structure: more advanced analysis schemes

M. A. PEDDER
University of Reading

In the previous two chapters, I have described some principles and applications of simple objective analysis methods suitable for estimating important dynamical and kinematic quantities from routinely available surface and upper air data. As an example, the least-squares-plane method was shown to provide useful measurements of several important atmospheric processes, yet could be applied to real data without the need for sophisticated computing facilities.

For more advanced studies of atmospheric structure, and for the needs of numerical forecasting, more powerful analysis techniques must be used. For any given problem, the final analysis must of course describe sufficient structure in the surface corresponding to the spatial variation of an observable quantity as is required to calculate the dynamical or kinematic properties which are related to that observable quantity. The essence of a good analysis scheme is not only that it should be able to provide a proper representation of this structure, but also that it should be able to cope with the constraints of variable data quality and spatial coverage. In this chapter I shall attempt to describe some of the principles used in modern objective analysis schemes and, where appropriate, compare the functions of such schemes with the more traditional techniques of chart analysis.

From plane to polynomial

The least-squares-plane surface representation of the local variation of, say, geopotential height (h) of a pressure surface in x, y map space (see equation 5.2; Pedder, 1981a) can be described mathematically as a 'truncated polynomial representation', and corresponds to a fit to the first three terms in a more general mathematical model surface given by

$$h(x, y) = a_0 + a_1 x + a_2 y + a_3 x^2 + a_4 xy + a_5 y^2 + a_6 x^3 \dots \text{etc.} \qquad (7.1)$$

where each 'a' coefficient is associated with a 'base function' in x and y which is simply of the form $x^m y^n (m \geq 0, n \geq 0)$. (The difference between the type of h

surface which can be represented by equation 7.1 and that obtained from the plane surface fit is similar to the difference between figures 5.1a and 5.1b in Pedder, 1981a.) Terms for which $m + n = k$, where k is a constant number, are said to be 'of order k'. Thus a_0 corresponds to terms order zero, a_1, a_2 to terms order 1, a_3, a_4, a_5 to terms order 2, and so on. The least-squares-plane analysis is thus 'of order 1'. It was shown earlier how this fit to observations provides a means of estimating the gradient properties $\partial h/\partial x$ and $\partial h/\partial y$ (see Pedder, 1981a). Such an analysis does not, however, provide a local estimation of a higher-order derivative property such as $\partial^2 h/\partial x^2$. To do this, we need to include terms of order 2 from equation 7.1 in the fitting equation. An example of where such an analysis model is required is when relative vorticity is to be estimated from the local geopotential height distribution. This 'geostrophic' relative vorticity (ζ_g) is related mathematically to height (h) and x, y co-ordinates by

$$\zeta_g = \frac{\partial v_g}{\partial x} - \frac{\partial u_g}{\partial y} = \frac{g}{f}\left(\frac{\partial^2 h}{\partial x^2} + \frac{\partial^2 h}{\partial y^2}\right), \qquad (7.2)$$

where f is the local Coriolis parameter and g, acceleration due to gravity. If h is fitted to a polynomial surface consisting of the first six terms in equation 7.1 (known in the following as a 'quadratic surface fit'), then the quantities $\partial^2 h/\partial x^2$ and $\partial^2 h/\partial y^2$ are then given by $2a_3$ and $2a_5$ respectively. Thus, the quadratic surface fit can provide estimations of the local values of dynamical or kinematic quantities which are not available directly from the least-squares-plane analysis. The quadratic surface analysis has been frequently used in research studies, the same principles being applied in the estimation of the six a coefficients as were described in Pedder (1981a) for the three least-squares-plane coefficients. However, for a quadratic fit, at least six station observations are required to secure one solution and, in order to avoid large sampling errors, more like ten station observations might be used in practice. This puts such an analysis scheme outside the scope of rapid hand calculation, though the solution procedure is quite trivial for a programmable computer.

Even this quadratic polynomial surface fit is not of a high enough order to provide estimations of some important dynamical processes. In Pedder (1981b), I mentioned the process of 'horizontal variation of thermal advection by the geostrophic wind field', which is one process that is closely associated with the synoptic development in mid-latitudes. A quantitative measure of this process requires estimations of quantities such as $\dfrac{\partial}{\partial x}\left(\dfrac{\partial}{\partial x}\left(\dfrac{\partial h}{\partial y}\right)\right)$, which can be obtained from a polynomial model surface only if terms of order 3 are included. Such a surface requires the solution of 10 coefficients, and may involve taking data from about 20 stations before one reliable estimation of the thermal advection process is obtained! Partly for this reason, high-order polynomial surfaces are not often used for large-scale analysis of meteorological fields, though this technique was one of the earliest methods of objective map analysis, having been introduced into meteorology by Panofsky (1949) more than thirty years ago.

Grid-point interpolation

An alternative to estimating a quantity such as $\partial h/\partial x$ directly from the coefficient of x in the least-squares-plane fit of h on x, y map space is to use the equation relating h to x and y as a means of *interpolating* the height data onto regularly spaced grid points. This corresponds to a procedure equivalent to the first form of objective analysis defined in Pedder (1981a). The interpolation is effected simply by substituting the x, y co-ordinates of each grid point into the estimated fitting equation

$$\hat{h}(x, y) = \hat{a}_0 + \hat{a}_1 x + \hat{a}_2 y. \tag{7.3}$$

If d is the distance between consecutive grid points lying along the x direction, $\partial h/\partial x$ at the point x, y can then be estimated as

$$\left(\frac{\widehat{\partial h}}{\partial x}\right)_{x, y} = \frac{\hat{h}(x + d, y) - \hat{h}(x - d, y)}{2d}, \tag{7.4}$$

where the quantities in parentheses following h indicate the position of the grid points being used. Equation 7.4 is an example of a 'finite difference' estimation which is described by Harwood (1981), and is discussed further in relation to numerical forecasting by Gadd (1981). Of course, such an approach is something of a waste of time if used merely in place of the estimation $(\partial h/\partial x) = \hat{a}_1$. However, if the interpolation is carried out at a number of different grid points using *different* groups of station observations at each one, then the coefficient $\hat{a}_1$ will generally vary from one grid point to another, which means that higher-order structure is being represented in the grid-point height field analysis than would be otherwise obtained if all the available observations were used to solve for just one plane surface. The way in which such a scheme might be applied in practice is represented schematically in figure 7.1.

One possible advantage of the grid-point interpolation method over the higher-order polynomial surface analysis is that the degree of smoothing of errors or small-scale fluctuations can be varied by varying the distance between grid points or by averaging grid-point estimations over several adjacent grid points, whereas in the case of the straightforward polynomial analysis this smoothing is fixed by the order of the polynomial model used and the spatial distribution of the observations. In grid-point differencing schemes, the magnitude of the gradient and higher-order derivative quantities tends to decrease on average as the distance between grid points is increased, due to the fact that the grid point analysis is not 'resolving' spatial structures which are not much larger than the distance between the grid points. Obviously, if the grid-point spacing is increased too much, then even the important synoptic-scale structures are not represented correctly in the finite difference estimations. In practice, the analyst must therefore select a grid-point separation which is small enough to resolve the structure of interest to him, without having it so small that his estimations become very sensitive to smaller-scale fluctuations. Thus, although the degree of smoothing *can* be adjusted by altering grid-point separation, the latter is often fixed through consideration of other factors, in which case the efficiency of the analysis is

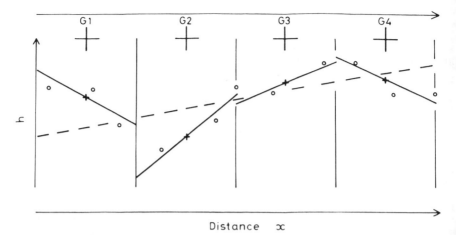

Distance x

Fig. 7.1 A schematic (one-dimensional) representation of how several fitted straight-line segments might be used to build up a grid point (G) field of the interpolated quantity, h. Three observations (marked ○) per segment are used to estimate each grid-point value (marked +). Within each segment the quantity $\partial h/\partial x$, i.e. the slope of the straight-line segment, is constant, but varies from one grid point to another. The dashed line represents the result of fitting all the observations to one straight line segment, in which case $\partial h/\partial x$ is the same at all the grid points.

ultimately determined by the original interpolation scheme. In this respect, polynomial surface interpolation is often not a very effective method, especially in regions where data are sparse, i.e. where the distance between adjacent observing stations is large compared with the distance between adjacent grid points.

Linear interpolation

The use of a polynomial model for carrying out interpolation of data onto grid points in map space is just one example of a large class of so-called 'linear interpolation' schemes. In general, if $\hat{q}$ is the estimated grid point value of q and q'_1, $q'_2 \ldots q'_m$ are observations of q, then a linear interpolation scheme is defined by the equation

$$\hat{q} = \sum_{i=1}^{m} w_i q'_i, \tag{7.5}$$

where w_i is the weight given to the ith observation (see equation 5.1; Pedder, 1981a). For a fixed set of m stations, the value of the weights may be calculated in various ways. The least-squares polynomial interpolation model provides one method for calculating these weights, but they could also be calculated more directly from station position data. One obvious method is to calculate w_i from an expression of the form

$$w_i = f(r_i), \tag{7.6}$$

where $f(r_i)$ is some simple analytical function of the distance r_i between the ith station and the grid point. Not all functions would be suitable, since we would normally require w_i to decrease with increasing distance r_i, but not to become infinitely large near $r_i = 0$. A further 'constraint' on the choice of weights might be that their sum should be unity, so that if $q'_1 = q'_2 = \ldots = q'_m$, then $\hat{q}$ takes on the same constant value. (A set of weights which satisfies this condition is said to be 'normalized'.) An example of a function which satisfies all these requirements is

$$w_i = S \times (d - r_i)^2 \text{ for } r_i < d, \tag{7.7}$$
$$w_i = 0 \qquad\qquad \text{ for } r_i \geqslant d,$$

where d is a constant, and S a normalizing constant given by

but

$$S = \Bigg/ \left[\sum_{i=1}^{m} (d - r_i^2) \right]^{-1},$$

where m is the number of stations falling within a distance d of the grid point. The set of weights obtained from such a model will depend not only on the positions of the m stations, but also on the value of d, which might be chosen quite arbitrarily such that m is limited to a 'reasonable' number, e.g. not more than 10. There are moreover a large number of other functions which would also satisfy these conditions, and it would appear at this point that the choice of 'interpolation model' is quite arbitrary.

How then does the choice of interpolation model affect the final analysis? Suppose that one model uses a function of the distance r_i which decreases very rapidly with increasing r_i, while a second uses a function which decreases very slowly with increasing r_i. Having calculated the normalized set of weights for both models applied to the same set of m stations we might arrive at the values as represented in figure 7.2. We notice that, using the first model, the grid point value $\hat{q}_0$ would depend mainly on the values of q observed at stations close to the grid point, whereas using the second model all the observations contribute significantly to the interpolated grid-point value, which would then turn out to be not much different from the arithmetic mean of all the observations. It follows that the second model would tend to smooth out small-scale fluctuations and random errors of observations, whereas the first model would produce relatively little smoothing. It is easy to see that both the grid-point-value inter-polated field and its derivative structure do not just depend on the data but are also 'model-dependent'. The problem of selecting an interpolation model which does just the 'right amount of smoothing', so as to retain the resolvable informa-tion represented in the data as much as possible whilst at the same time rejecting the unresolvable small-scale structure and random errors, is considered in a later

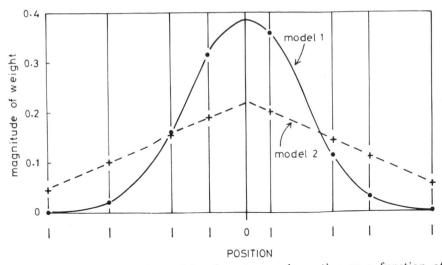

Fig. 7.2 Magnitude of the weight given to an observation as a function of distance from the point of interpolation (marked o). Positions of eight stations are indicated by the marks along the position axis. Model 1 corresponds to a strongly varying function of distance, model 2 to a weakly varying function of distance.

section. But first, it is necessary to introduce a slight modification to the original linear interpolation model represented by equation 7.5.

Background fields

Most practical interpolation schemes used in numerical forecasting differ from that represented by equation 7.5 in that the observations $q'_1, q'_2 \ldots q'_m$ are replaced by their departure from some 'background' or 'first guess' value. One obvious choice of background field is that represented by the climatological distribution of q appropriate to the time of the analysis. The linear interpolation of deviations from the climatological field is then given by

$$(\hat{q}_0 - \bar{q}_0^{\infty}) = \sum_{i=1}^{m} w_i(q'_i - \bar{q}_i^{\infty}), \tag{7.8}$$

where the overbar $^{-\infty}$ stands for the climatological mean value. Suppose that, on one occasion, all the m observed values of q happen to be the same as their climatological values, then the right-hand side of equation 7.8 would be zero, so that $\hat{q}_0 - \bar{q}_0^{\infty} = 0$ or $\hat{q}_0 = \bar{q}_0^{\infty}$. This is clearly a reasonable value to give $\hat{q}_0$, but would not generally result from using the scheme represented by equation 7.5 unless $\bar{q}_0^{\infty} = \bar{q}_1^{\infty} = \bar{q}_2^{\infty} = \ldots = \bar{q}_m^{\infty}$ (which is obviously unlikely). Moreover, if the condition that the weights $w_1, w_2 \ldots w_m$ should be normalized is relaxed, then, when the observing stations are all at a great distance from the grid point, the interpolation value $\hat{q}_0$ will tend to be close to $\overline{q_0}^\infty$, which would again seem to be the best estimation in the absence of data. Thus, the use of a climatological background field can be seen to improve the 'reality' of map analysis, particularly in data-sparse regions. Of course, with this scheme there is the problem of knowing what is the climatological value of q at all grid points as well as all observing points although, since the climatological fields of the common meteorological observables tend to be 'smooth', this need not present much of a problem in practice.

Although climatological background fields have been used in objective analysis schemes, many modern analysis schemes use forecast or 'prediction' background fields instead. Thus, the predicted values of q given by some numerical forecast are used in place of the climatological mean quantities in equation 7.8. The difference between the predicted and the true value of q at any point is then known as 'the prediction error'. This procedure has an obvious advantage over the use of climatological background values in data-sparse regions, provided that the forecast model produces better predictions for q than that given by climatology (which is now generally the case for short-term forecasts). In such a scheme, the observations are thus seen to provide a *correction* to the forecast field, this correction being based on the observed prediction errors. If forecast models were perfect there would, of course, be no need for observations at all, since the observed prediction errors at all stations would differ from zero only by an amount equal to the random observing error, and the right-hand side of equation 7.5 would therefore be virtually zero when m is large. There are, however, good theoretical reasons for supposing that such perfect predictions are not realizable, so we are unlikely to be in the position of not requiring observations!

Statistical interpolation: the 'optimal' solution

All the linear analysis schemes described so far have one feature in common, namely that the weight given to an observation depends primarily on the position of that observation relative to the point of interpolation, this dependence being determined by an arbitrarily chosen weighting model. In 1963, the Russian meteorologist Gandin suggested an alternative approach to weighting observations which would not only avoid the arbitrary aspects of other weighting schemes, but, in principle, could also lead to the best possible analysis in the sense that the total interpolation error would be, on average, a minimum. The essence of Gandin's 'optimal interpolation' method is that the weighting given to observations should depend primarily on the relationship between the observations themselves rather than the spatial relationship between the points of observation. The arguments put forward by Gandin may only be fully appreciated by those familiar with statistical theory, but some idea of how this method works might be described without the mathematics as follows:

Suppose we collect observations of surface pressure from two stations A and B at which we know the climatological 'background' pressures $\overline{p}_A^\infty$ and $\overline{p}_B^\infty$. Now, consider the differences between simultaneously observed pressures p_A, p_B and their climatological values – call these differences p'_A and p'_B, so that $p'_A = 0$ when $p_A = \bar{p}_A^{\infty}$, and $p'_B = 0$ when $p_B = \bar{p}_B^{\infty}$. If we find that whenever p'_A is large and positive (or negative), p'_B is also large and positive (or negative), then we can conclude that the observations at A and B are strongly *correlated*. Therefore, if, on one occasion, no observation were made at station A, then we might reasonably substitute the *estimation* $\hat{p}'_A = p'_B$, or $\hat{p}_A = \bar{p}_A^{\infty} + p'_B$, in place of the missing observation. On the other hand, if there were no obvious correlation between p'_A and p'_B on average, then the best estimate we could make would probably be $\hat{p}_A = \bar{p}_A^{\infty}$, that is, set p_A equal to its climatological mean value regardless of the observation at B. Notice that the difference between these two extremes is that, in the first case, we have given unit weighting to the quantity p'_B, but, in the second case, zero weighting. Thus the weighting has been made dependent on the degree of similarity between the deviation quantities p'_A and p'_B. This 'similarity' can be described quantitatively with reference to the statistical measurement $\overline{p'_A p'_B}^{\infty}$ – the average product of the deviations as estimated from a large number of simultaneous observations – known as the *covariance* between p_A and p_B. If there were no correlation between the observations at A and B, then this covariance quantity would be zero, whereas if they were highly correlated, then the covariance would generally be large and positive.

Gandin developed a linear weighting model in which the weights given to various observations depend on station-to-station and station-to-grid point covariances, such that the linear interpolation based on these weights gave the smallest average interpolation error, when compared against any other possible choice of the weighting factors. The covariance between grid point values and observations at a particular station cannot, of course, be determined directly, but this statistic can be assumed roughly equal to the value corresponding to two observing stations which are separated by about the same distance.

One important advantage of this scheme is that it can cope easily with observations from different sources, even when the quality of observation depends on the source. Thus a satellite-derived temperature at a point 500 km from a certain grid point would be given less weight than a radiosonde-derived temperature 500 km from the same grid point, this being because radiosonde temperatures are at present more reliable than satellite-derived temperatures.

Although Gandin developed this optimal interpolation scheme using deviations from climatological background fields, the same approach may be applied to deviations from other types of background fields. Thus, the method may also be applied to deviations from predicted values of an observable quantity – the 'prediction errors' as defined in the previous section – and this approach is now being widely adopted for global analysis of meteorological data in what are now more generally described as 'statistical' interpolation schemes.

Although this statistical approach to meteorological analysis apparently offers the best solution to the problem of extracting as much useful information as possible from the available observations, it is nevertheless a complicated scheme to apply in practice, and generally requires more computer effort than the simpler distance-weighting methods. There are also considerable problems associated with establishing accurate covariance estimations, which are essential for the 'correct' choice of weighting in the linear interpolation model. Partly for these reasons, the statistical approach to data analysis has only been widely adopted for numerical forecasting during the last five years or so, more than twenty years after the basic theory was first established.

Multivariate analysis

So far, I have mainly considered the problem of interpolating the value of one meteorological variable such as geopotential height onto a grid point in horizontal map space using a linear combination of observations of the same variable at a number of surrounding stations. In practice, a more convincing analysis is often obtained if the interpolated grid-point field is forced to be physically or dynamically consistent with the interpolated field of some other variable. This involves carrying out an analysis using a linear combination of two or more observed variables together, the spatial relationship between them being 'built' into the analysis model. Such a procedure is known as a 'multivariate analysis scheme' when effected numerically by a computer, though the concept of multivariate analysis is almost as old as the art of chart analysis itself.

An obvious example of multivariate analysis as applied to mid-latitude data is to be found in the use of upper wind observations as an aid to the subjective analysis of an upper-level height map: the analyst uses station observations of wind as a guide for the direction of his analysed height contours, while the spacing between consecutive contours can be checked against the observed wind speed using a geostrophic scale, which is often printed on the blank chart. These 'rules of thumb' are based, of course, on the observation that, in middle and high latitudes, the wind and pressure fields are always nearly in geostrophic balance, which can be expressed mathematically by the equations relating u and v wind components to the geopotential height, h, of a pressure surface:

$$u \simeq -\frac{g}{f} \frac{\partial h}{\partial y}, \tag{7.9a}$$

$$v \simeq \frac{g}{f} \frac{\partial h}{\partial x}, \tag{7.9b}$$

where $g = 9.80665$ m s^{-2} and f is the local value of the Coriolis parameter (about 1.1×10^{-4} s^{-1} at 50°N). Having expressed this relationship mathematically, it is fairly easy to incorporate into a numerical scheme. As a simple example, consider again the least-squares-plane analysis of height h on x, y map space which results in the fitted height surface

$$\hat{h}(x, y) = \hat{a}_0 + \hat{a}_1 x + \hat{a}_2 y. \tag{7.10}$$

Differentiating this equation with respect to x and multiplying by g/f gives, with equation 7.9b,

$$\frac{g}{f} \frac{\partial \hat{h}}{\partial x} = \frac{g \hat{a}_1}{f} = \hat{v}. \tag{7.11a}$$

Similarly, after differentiating with respect to y,

$$\frac{g}{f}\frac{\partial \hat{h}}{\partial y} = \frac{g\hat{a}_2}{f} = -\hat{u}. \tag{7.11b}$$

Upper air data at one station can provide observed values for h, u and v, which are sufficient to solve for all three coefficients in the three equations 7.10, 7.11a and 7.11b, whereas using the height data alone requires data taken from at least three stations. It also follows that, if we use both height *and* wind data from three stations, we could solve for up to nine coefficients in a polynomial surface fit as described in the first section of this chapter. A quadratic polynomial surface fit requires the estimation of only six coefficients, so that three stations can provide more than enough data to solve for this higher-order surface fit. An example of such an analysis is shown in figure 7.3, where it is compared with the plane fit given from solving equation 7.10 using the height data alone. I leave it to the reader to decide which is the more convincing analysis.

A second important constraint on upper level height analysis is that the geopotential height (h) of a pressure surface (p) is everywhere consistent with both the mean temperature ($\bar{T}$) of the atmosphere between mean-sea-level (msl) and the pressure level p and the local value of msl pressure (p_0). This follows from the observation that the vertical pressure-gradient force almost exactly balances the gravitational force in our atmosphere (see Panofsky, 1981). The relationship between h, p, p_0 and $\bar{T}$ can be expressed mathematically in the barometric equation

$$h = \frac{R\bar{T}}{g} \ln (p_0/p), \tag{7.12}$$

where R is the specific gas constant for air. This relationship is again applied in subjective chart analysis in the form of the technique known as 'gridding', often used as an aid in the analysis of the 1000–500 mb thickness chart. Here, the analyst produces charts of 500 mb height (involving the 'multivariate' assimilation of 500 mb winds) and msl pressure. The 500 mb chart is then laid over the msl isobaric analysis and the 1000–500 mb thickness at a number of arbitrarily chosen points calculated as the difference between the interpolated 500 mb height and the height of the 1000 mb surface above msl, the latter being estimated, often using a suitable nomogram, as a function of msl pressure and surface temperature. The gridding principle can again be built into numerical analysis models by analysing simultaneously height, surface pressure and temperature data.

Apart from these types of constraints, in practice, it is also necessary to establish other more subtle relationships between analysed pressure (or height), temperature and wind fields, when the grid-point fields are to be used for numerical prediction. However, such constraints, which typically control the type of atmospheric wave structure that is allowed to develop during numerical integrations, cannot be incorporated into the linear analysis schemes described here. These 'initialization' procedures must be carried out separately, and bear no resemblance to procedures familiar to the human analyst.

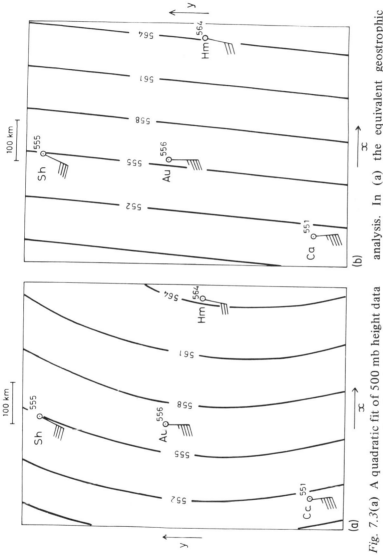

Fig. 7.3(a) A quadratic fit of 500 mb height data on x, y map space based on height and wind data from three U.K. aerological stations compared with (b) a plane fit based on height data alone, station Au not being used for either analysis. In (a) the equivalent geostrophic wind speed varies from 50 kts in the extreme west to 35 kts in the extreme east. In (b) the equivalent geostrophic wind speed is everywhere 45 kts.

Man and machine

Within the contexts of numerical forecasting and large-scale meteorological studies, objective analysis procedures have largely replaced more traditional subjective chart analysis methods. Nevertheless, this does not mean that the machine product is always better than that produced by the expert human analyst. Indeed, a subjective map analysis is still often used as a standard against which to assess the map produced by the computer program, and there is little doubt that useful dynamical and kinematic measurements *can* be obtained from subjectively analysed data. It is also interesting to notice how many of the ingredients of objective analysis are, in practice, numerical analogues of procedures familiar to the expert human analyst. Examples of the application of geostrophic and hydrostatic constraints in multivariate analysis were described in the last section, but more subtle analogues can also be found, e.g. the adjustment of the forecast background field by linear interpolation of prediction error is similar to the principle of 'time continuity' in subjective chart analysis, which involves using a previous analysis as a basis for the latest analysis, having first moved significant meteorological features such as fronts and depression centres (using empirical forecasting rules) by distances consistent with the time difference between analysis charts. Even the basic structure of linear interpolation itself is in some ways analogous to the human operation – the analyst does not draw isobars on a surface map by looking at one observation at a time, but rather by scanning all the observations within a reasonable area of the chart, 'weighting' them according to whether they are close to or distant from the current pencil position, and only then drawing an isobar segment.

As is true of many machine analogues of human activities, there are, however, some aspects of meteorological analysis which are familiar to the synoptic meteorologist, but which are very difficult to program in a numerical form. Because of this, the human analyst may on some occasions produce the more realistic, and probably the more useful analysis. An important example is in the treatment of the 'anomalous' observation: at present only the experienced meteorologist is capable of distinguishing, with reasonable certainty, between what might be a large observing or reporting error, and a genuine but anomalous observation. This is largely because the experience and wisdom required to make such a judgment clearly cannot easily be built into a computer program. This ingredient of subjective analysis is particularly important when dealing with data-sparse regions, especially in the early detection of relatively small-scale features which may eventually develop to become synoptically important weather phenomena (the detection and analysis of 'polar lows' in northerly outbreaks over the north-eastern Atlantic is a typical example). On the other hand, it is important to appreciate that, for the purposes of estimating dynamical and kinematic quantities, it is doubtful whether the results obtained from a subjective analysis can ever be considered as quantitatively reliable as those obtained from an objective method. Even then results may only be considered reliable where there is a reasonably good coverage of high quality observations. Even with the enormous increase in certain types of data coverage through the use of meteorological satellites, it is probably still the case that reliable estimations of dynamical and kinematic processes (as, for example, described in Pedder, 1981b) are possible only within less than 20 per cent of the total atmospheric volume below 20 km, where observations are of sufficient density. In many ways, it is still very much easier to describe our atmosphere with the results of theoretical models than it is to describe it with the results obtained from practical measurement!

References

Gadd, A. J. (1981) 'Numerical modelling of the atmosphere', this volume, 194–204.

Gandin, L. S. (1963) *Objective Analysis of Meteorological Fields*, translated from Russian by the Israeli Program for Scientific Translations 1965.

Harwood, R. S. (1981) 'Atmospheric vorticity and divergence', this volume, 33–54.

Panofsky, H. A. (1949) 'Objective weather map analysis', *J. Met.*, 6, 386–92.

Panofsky, H. A. (1981) 'Atmospheric hydrodynamics', this volume, 8–20.

Pedder, M. A. (1981a) 'Practical analysis of dynamical and kinematic structure: principles, practice and errors', this volume, 55–68.

Pedder, M. A. (1981b) 'Practical analysis of dynamical and kinematic structure: some applications and a case study', this volume, 69–86.

8
Atmospheric waves

B. W. ATKINSON
Queen Mary College,
University of London

As in all natural systems, the state of the atmosphere, including the configuration of its circulation, is a function of processes operating through time. Any particular circulation results from a particular combination of processes. As we know that atmospheric processes are well described by the equations of motion, the continuity equation, the thermodynamic or heat equation, the moisture equation and the equation of state (see Panofsky, 1981), this means that the dominance of particular terms (e.g. gravity or Coriolis effects) in these equations will probably lead to certain types of atmospheric circulation. Clearly, a deeper understanding of atmospheric behaviour ensues from knowing which terms produce which circulations: in turn, understanding sometimes leads to better prediction.

Another look at the equations of motion

Before we can analyse the effects of the various forces upon circulations we need to know a little more about some of the basic equations that describe the processes. Panofsky showed that the acceleration of air was due to the combined effects of the pressure-gradient force, the geostrophic or Coriolis force and the frictional force. If we concern ourselves with the 'free' atmosphere we can simplify the problem by ignoring the frictional force. The equation of motion in the zonal (x) direction may then be written as

$$\frac{du}{dt} = fv - \frac{1}{\rho} \frac{\partial p}{\partial x},$$

(8.1)

where the left-hand side represents the rate of change with time of the zonal velocity u of a parcel of air, the first term on the right-hand side represents the Coriolis effect and the second term the horizontal pressure-gradient force in the x direction. In the Coriolis term, f is the Coriolis parameter, or the vertical component of the earth's vorticity, $2\omega \sin \phi$ (ω – angular velocity of the earth's rotation; ϕ – latitude), and v is the air velocity in the meridional (y) direction. In the last term, ρ is the air density and p is the air pressure. The left-hand side of equation 8.1 can be expanded to identify the changes of u not only intrinsically with time (i.e. changes that would occur even if the air particle did not *move*) but also those that occur because it moves through *space* over a period of time. Thus, the differential operator d/dt can be expanded into the following partial differentials:

$$\partial/\partial t + u\partial/\partial x + v\partial/\partial y + w\partial/\partial z.$$

The last three operators describe the changes due to moving in each of the three directions x, y, z where z is the vertical. Now equation 8.1 may be rewritten as:

$$\frac{\partial u}{\partial t} + u\frac{\partial u}{\partial x} + v\frac{\partial u}{\partial y} + w\frac{\partial u}{\partial z} = fv - \frac{1}{\rho}\frac{\partial p}{\partial x}. \tag{8.2}$$

This is a partial differential equation, known as such because it contains more than one independent variable (i.e. x, y, z and t). We may now decide whether equation 8.2 is *linear*. In a linear equation, no term contains the product of two *dependent* variables (i.e. u, v, w, ρ, p), the square or higher power of a dependent variable, the product of two derivatives or the product of two dependent variables. Clearly, equation 8.2 is *non-linear*. This is an important characteristic because very little is known about the properties of solutions of non-linear equations, whereas the opposite is true of solutions to linear equations. These general observations on the nature of equation 8.2 also apply to the equations of motion for the meridional component of velocity (v) and the vertical velocity (w).

As outlined in the introduction one of the main aims of 'solving' equations such as equation 8.2 is to establish both configurations of airflow (in terms of u, v, w) that may occur in our fluid atmosphere and, even more important, the basic causes of those configurations. The aim is primarily to achieve a general understanding of the nature of the airflow rather than predicting its configuration at a particular time and place. We noted in the introductory chapter of this book (Atkinson, 1981) that the equations of motion (e.g. equation 8.1) contained accelerations, i.e. time derivatives of velocities, which means that their 'solutions' will principally be in terms of those velocities. Thus, to see any 'solution' as relevant to familiar patterns such as Rossby waves on upper air maps, we must look at the waves on our map in terms of their velocity structure (figure 8.1). This mental jump is fundamental to an appreciation of what dynamical meteorology is all about. Once the velocities are known, one can, of course, calculate displacements and trajectories.

There is an attractive simplicity to the conceptual framework outlined above: indeed it could encourage the uninitiated to believe that all meteorological problems should have been solved decades ago. As is so often the case however, the meteorological problems which are 'easy' to formulate are frequently far from easy to solve.

The solution of the meteorological equations is particularly difficult because they are non-linear. Such solutions as are available have resulted from two basic approaches; analytical and numerical. The analytical methods were, of course, used before the numerical, providing the bulk of dynamical meteorological theory

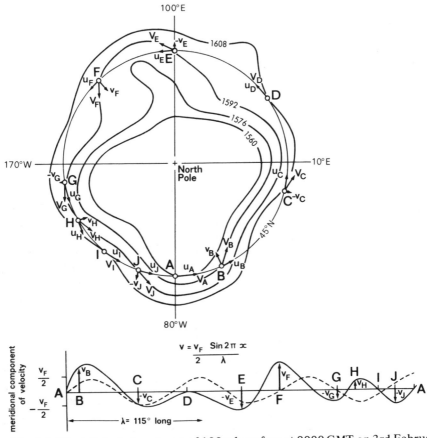

Fig. 8.1 (Top) Selected contours of 100 mb surface at 0000 GMT on 3rd February 1980 showing wavelike flow. Where a contour (equivalent to a streamline) intersects latitude $45°N$ the velocity (V) has been resolved into zonal (u) and meridional (v) components. In the latter, positive values are poleward. (Bottom) For all points A to J, the v component has been plotted against longitude, showing a wave form. For comparison a sinusoidal wave (dashed line) has also been plotted, representing $v = (v_F/2) \sin \dfrac{2\pi}{\lambda} x$, where x is the distance along latitude $45°N$.

up to about 1960. Since that date, numerical modelling has become a major research technique and the basic methods are outlined by Gadd (1981). The remainder of this chapter is concerned with some important results achieved by one type of analytical work — work principally undertaken between the years 1930 and 1960. Analytical work as a whole is of course still an active field of study.

Perturbation method

Analytical solutions of the meteorological equations were largely achieved by the use of the *perturbation method*. In essence, this method allows the linearization of the otherwise non-linear equations and the derivation of 'wave' solutions. By the latter, we mean that a 'wave-shaped' configuration of a dependent variable satisfies the linearized equations. Substitution of the wave solutions into further equations representing boundary conditions (such as vertical velocities being zero on a horizontal earth surface) results in a form of 'dispersion equation'. This type of equation relates the speed of the wave to its wavelength as well as the physical parameters of the problem. The dispersion equations are valuable because they allow the wave speed to be computed from the wavelength and the mean zonal wind and they also allow the stability of the wave to be determined. Within this general technique, it is the specification of the certain conditions which determines the character of the waves. Thus, if we specify that gravity is to be considered, but the rotation of the earth is to be ignored, then clearly waves resulting primarily from the effects of gravity will be revealed by the analysis. Nevertheless, this result is of significance as it tells us which atmospheric circulations are primarily caused by the action of gravity – fulfilling the aims outlined in the introduction to this chapter.

Before outlining the perturbation method it will help to know how waves in general are described mathematically and also some types of waves that could exist in the atmosphere. The mathematical description of the waves to be considered assumes constant shape and constant velocity of propagation.

Mathematical description of waves

Figure 8.2 shows the positions of a typical wave, such as would occur as a result of 'flicking' a length of rope, at two times, t_1 and t_2. To simplify the exposition we assume that the wave shape remains constant. The wave velocity is c. The shape of the wave profile at time t_1 could be described by the function

$$n = f(x).$$

Turning attention to the wave at time t_2, the point $0'$ on the x-axis is in the same position relative to the t_2-profile as the point 0 was to the profile at time

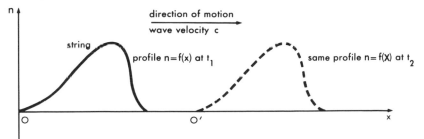

Fig. 8.2 Waves along a string (after Richards and Williams, 1972).

t_1. Let the distance along the x-axis referred to $0'$ be measured by the quantity X. Thus the shape of the profile referred to $0'$ at time t_2 is

$$n = f(X).$$

For the illustration of wave parameters it is of course inconvenient to have the origin of co-ordinates moving along with the wave profile; we wish to refer the profile to the fixed point 0. Now, since the profile is moving with constant velocity, c, the distance $00'$ is $c(t_2 - t_1)$. Thus,

$$X = x - c(t_2 - t_1).$$

The profile at time t_2 is thus described by

$$n = f[x - c(t_2 - t_1)].$$

As we can set $t_1 = 0$ without losing generality the equation reduces to

$$n = f(x - ct), \tag{8.3}$$

where t is now any time after the wave passed the origin.

Given the function f, equation 8.3 completely defines a one-dimensional wave of constant profile moving with a constant velocity c along the positive direction of the x-axis. For a wave that is similar in all respects, but moving in the opposite direction, the equation would be

$$n = f(x + ct).$$

Note that n could represent parameters such as displacements and velocities (see figure 8.1).

So far, the form of the function f in equation 8.3 has been unspecified; i.e. the wave profile may have the shape of any continuous curve. In fact, the simplest types of wave to treat analytically are those whose profiles are pure sine or cosine waves, an example of the former being

$$n = f(x - ct) = a \sin b(x - ct), \tag{8.4}$$

where a and b are constants. If we use equation 8.4 to plot the value of n against x for a given value of t, the curve in figure 8.3 results. Since the sine function is periodic, the wave profile repeats itself after fixed intervals of x. The repeat distance is known as the wavelength and is usually designated by λ. If we increase x by λ in equation 8.4 the values of n will, by definition, be unaltered.

$$n = a \sin b(x - ct) = a \sin b((x + \lambda) - ct).$$

The smallest quantity we can add into the phase of the sine function leaving it unaltered for all values of x is $360°$ or 2π. Hence

$$b\lambda = 2\pi \quad \text{or} \quad b = \frac{2\pi}{\lambda}.$$

Thus, we now see that a is the amplitude of the wave and b takes the form $2\pi/\lambda$, where λ is wavelength. As waves whose profiles are those of a cosine func-

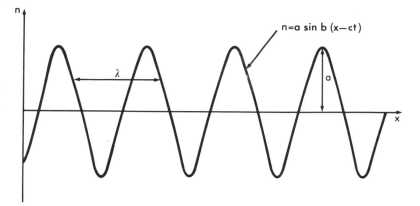

Fig. 8.3 Graph of the function $n = a \sin b(x - ct)$ as shown in equation 8.4 (after Richards and Williams, 1972).

tion are very similar to those with sine function (the former being simply the latter shifted by a one quarter wavelength), we may write

$$n = A \sin \frac{2\pi}{\lambda} (x - ct), \qquad (8.5)$$

or

$$n = B \cos \frac{2\pi}{\lambda} (x - ct). \qquad (8.6)$$

Either of these equations could, for example, describe a streamline of airflow, but the real flow patterns of the atmosphere are more likely to be represented by numerous sinusoidal waves superimposed on each other. A further slight modification to equations 8.5 and 8.6 may be the addition of a 'phase parameter', because the wave does not necessarily reach its maximum at $x = 0$, $t = 0$. Thus, x_0 in equation 8.7

$$n = B \cos \frac{2\pi}{\lambda} (x - ct - x_0) \qquad (8.7)$$

is the phase parameter. Its effect is to shift the position of the peak of the wave to x_0 when $t = 0$.

Further analysis of waves benefits from the re-writing of the trigonometric relations in equations 8.5 and 8.6 as exponential relationships. The basis of this transformation lies in the theory of complex numbers which tells us that

$$e^{i\theta} = \underset{\text{(real part)}}{\cos \theta} + \underset{\text{(imaginary part)}}{i \sin \theta}, \qquad (8.8)$$

where e is the base of natural logarithms (equal to 2.71828 . . .) and i is the imaginary quantity $\sqrt{-1}$. Manipulation of the above relationship gives us

$$\sin \theta = \frac{e^{i\theta} - e^{-i\theta}}{2i} \; ; \quad \cos \theta = \frac{e^{i\theta} + e^{-i\theta}}{2} \; .$$

Before applying this relationship to equation 8.7, we may re-write the latter as follows:

$$n = C \cos \frac{2\pi}{\lambda} (x - ct) + D \sin \frac{2\pi}{\lambda} (x - ct), \tag{8.9}$$

using a standard trigonometrical identity. On substituting exponentials, equation 8.9 becomes

$$n = C' e^{i(2\pi/\lambda)(x - ct)} + D' e^{-i(2\pi/\lambda)(x - ct)} \tag{8.10}$$

where $C' = \frac{1}{2}(C - iD)$ and $D' = \frac{1}{2}(A + iB)$. Equations 8.9 and 8.10 are equivalent expressions for sinusoidal waves. The advantage of expressing the waves in the form of equation 8.10 is that exponentials are much easier to handle mathematically than sines and cosines; they are easier to integrate, differentiate and sum as series. The procedure is thus as follows. We express our sine or cosine waves in exponential form; then we carry out our manipulation and take the imaginary or real part of the result, according to equation 8.8, as the quantity which is physically meaningful.

For certain values of the wavelength λ, the wave equation may yield a complex wave velocity c; such that

$$c = c_r + i c_i, \tag{8.11}$$

where c_r is the real part and c_i is the imaginary part of the wave velocity. Both c_r and c_i can only be real numbers. If c_i is zero, the wave velocity is a real number; if c_r is zero while c_i is not zero the wave velocity is an imaginary number. Substitution of equation 8.11 into equation 8.10 yields

$$n = C' e^{(2\pi/\lambda) c_i t} e^{i(2\pi/\lambda)(x - c_r t)} + D' e^{-(2\pi/\lambda) c_i t} e^{-i(2\pi/\lambda)(x - c_r t)}. \tag{8.12}$$

Equation 8.12 may seem a long way from our easily observed waves on a string, but it does in fact allow us to gain insights not only into those but also many other types of wave, including those in the atmosphere.

In equation 8.12, the terms $C' e^{(2\pi/\lambda) c_i t}$ and $D' e^{-(2\pi/\lambda) c_i t}$ are considered as amplitudes of the sinusoidal functions. If c_i turns out to be zero, these amplitudes are independent of time, and thus the waves keep the same amplitude. Such waves are called *neutral* or *stable* waves. On the other hand, if c_i is not zero, one of the terms in equation 8.12 will grow exponentially with time. Unless the boundary conditions are such as to make the amplitude of this term zero, the wave will grow and, by definition, be *unstable*. Thus, in summary, waves are stable when $c_i = 0$ and possibly not when c_i differs from zero. For unstable waves, c_i determines the rate of growth of the waves. In all cases, c_r measures the rate of progress of the wave in the x direction. Thus, the wave equation provides three kinds of information: the stability or otherwise of a wave; the degree of instability; and the speed of the wave.

Pure types of wave motion in the atmosphere

It is now appropriate to consider pure types of wave motion which may occur in the atmosphere. A convenient three-fold division is as follows; longitudinal, vertical-transverse and horizontal-transverse (figure 8.4). Longitudinal (or compression) waves are those in which the particle trajectories lie in the lines parallel to the direction of wave propagation. Sound waves are examples of this type. Vertical-transverse waves are those in which particles move up and down and back and forward while the waves are propagated horizontally. Gravity waves and, in particular, lee waves are examples of this type. Horizontal-transverse waves are those in which the particles move north and south (meridionally) while the waves are propagated along latitude circles (zonally). The planetary or Rossby waves are examples of this type.

The actual method

With our new appreciation of waves, mathematical and real, we are now in a position to see how perturbation methods yield useful results about the latter. The main point of the perturbation method is that it transforms non-linear differential equations into linear homogeneous equations. This is done as follows: (1) A simple, basic flow is chosen to satisfy the fundamental equations such as equation 8.2. (2) Small disturbances, or perturbations, are superimposed on this basic flow. The combined flow is assumed to satisfy the fundamental equations. (3) The perturbations are so small that products of perturbation quantities can be neglected with respect to terms of the first order in the perturbation quantities.

We can now apply these ideas to equation 8.2. To simplify the problem we further assume that the flow is incompressible, that is, density is constant for any fluid particle. The quantities relevant to the total flow are indicated by a bar, the undisturbed condition by a capital letter and the disturbance by a small letter. Thus,

$$\bar{u} = U + u. \tag{8.13}$$

For the total motion, the equation is

$$\frac{\partial \bar{u}}{\partial t} + \bar{u}\frac{\partial \bar{u}}{\partial x} + \bar{v}\frac{\partial \bar{u}}{\partial y} + \bar{w}\frac{\partial \bar{u}}{\partial z} = -\frac{1}{\rho}\frac{\partial \bar{p}}{\partial x} + f\bar{v} - 2\omega\bar{w}\cos\phi. \tag{8.14}$$

Substituting the undisturbed motion and the disturbance (equation 8.13)

$$\frac{\partial u}{\partial t} + \frac{\partial U}{\partial t} + (u+U)\frac{\partial u}{\partial x} + (u+U)\frac{\partial U}{\partial x} + (v+V)\frac{\partial u}{\partial y} + (v+V)\frac{\partial U}{\partial y} + (w+W)\frac{\partial U}{\partial Z}$$

$$+ (w+W)\frac{\partial u}{\partial z} = -\frac{1}{\rho}\frac{\partial p}{\partial x} - \frac{1}{\rho}\frac{\partial P}{\partial x} + fv + fV - 2\omega W\cos\phi - 2\omega w\cos\phi. \tag{8.15}$$

107

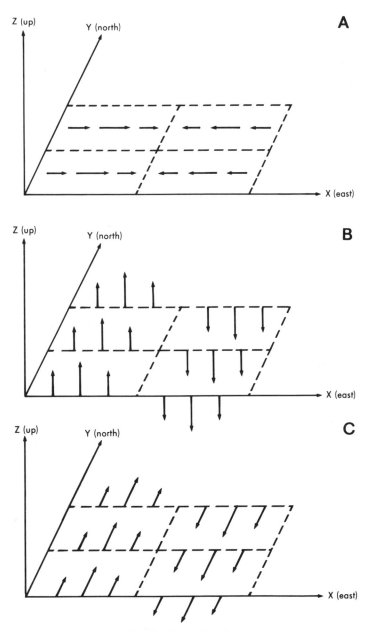

Fig. 8.4 Schematic distribution of velocity perturbations associated with (a) eastward-moving compression waves; (b) eastward-moving vertical-transverse waves; and (c) eastward-moving horizontal-transverse waves (after Thompson, 1961).

Similarly, the undisturbed motion is governed by

$$\frac{\partial U}{\partial t} + U\frac{\partial U}{\partial x} + V\frac{\partial U}{\partial y} + W\frac{\partial U}{\partial Z} = -\frac{1}{\rho}\frac{\partial P}{\partial x} + fV - 2\omega W \cos \phi. \qquad (8.16)$$

Subtracting equation 8.16 from equation 8.15 and neglecting terms containing products of the perturbation quantities gives

$$\frac{\partial u}{\partial t} + u\frac{\partial U}{\partial x} + U\frac{\partial u}{\partial x} + v\frac{\partial U}{\partial y} + V\frac{\partial u}{\partial y} + w\frac{\partial U}{\partial z} + W\frac{\partial u}{\partial z} = -\frac{1}{\rho}\frac{\partial p}{\partial x} + fv - 2\omega w \cos \phi. \qquad (8.17)$$

Since U, V and W are predetermined in the basic flow, this equation contains only terms of the first power in the unknown perturbation quantities u, v, w and p or their derivatives. This property makes the equation linear. It is also homogeneous as every term contains an unknown or its derivative. In fact, a solution giving waves travelling in the x direction will satisfy equation 8.17. It would of course be possible to derive linear equations for the v and w equations of motion which in turn would also have wave solutions. All the wave solutions would be useful only after they have satisfied certain *boundary conditions*. A typical, and easily appreciated, boundary condition is that the velocity component at right angles to the earth's surface shall be zero at all times. More subtle conditions may be used, such as the kinematic condition that the velocity components normal to an internal boundary (such as a front) must be the same as the two sides, otherwise a hole would form in the fluid. The application of boundary conditions to general solutions of the form of equation 8.12, where the dependent variable, therein called n, may in fact frequently be u, v, w or p, allows the derivation of relationships between wave speed (c), wavelength (λ), basic current speed (U say), and other parameters.

Results

We now turn to some of the results achieved by this method, looking in turn at sound waves, gravity waves, inertia waves, Rossby waves, and baroclinic waves.

Sound waves

Sound waves are longitudinal or compression waves. Analysis of them requires the first equation of motion (equation 8.2) *without* the Coriolis and frictional terms (the latter is excluded from equation 8.2 anyway), a relationship between density and pressure and the equation of continuity, including density variation (i.e. not incompressible). The dependent variables are perturbation pressure, perturbation density and a one-dimensional perturbation velocity (u). Transverse waves are excluded by the condition $v = w = 0$. A wave or periodic solution to the perturbation equations for pressure, density and velocity gives a wave speed (c) as

$$c = U \pm \left(\frac{c_p}{c_v} RT\right)^{1/2}. \qquad (8.18)$$

where c_p, c_v are specific heats at constant pressure and constant volume, R is the gas constant and T is the absolute temperature in the equilibrium state. For $T = 273$ K, $c = U \pm 331$ m s^{-1}.

Gravity waves

These are vertical-transverse waves whose primary driving force is that of gravity. Effects of compression, the earth's rotation and friction are ignored to isolate the mechanisms at work in pure gravity waves. Consequently, these terms are omitted from the u and w equations of motion and the two-dimensional continuity equation when they are used in this problem. Application of perturbation techniques again gives waves with the following speed:

$$c = U \pm \left(\frac{g\lambda}{2\pi} \tanh \frac{2\pi h}{\lambda} \right)^{1/2}. \tag{8.19}$$

where h is the height of the surface (or discontinuity between two uniform layers) on which the wave forms, and the other symbols have previously defined meanings. The wave velocity is thus made up of two terms: the first saying that the wave system is carried along by the current; and the second representing the effects of gravity as shown by the presence of g. Two extreme cases of equation 8.19 are useful. As we know that the value of $\tanh x$ approaches unity as x increases, then as the depth of h of the fluid increases relative to the wavelength, equation 8.19 becomes

$$c = U \pm \left(\frac{g\lambda}{2\pi} \right)^{1/2}. \tag{8.20}$$

This relation gives the velocity of waves in deep fluid. 'Deep' in this case means that h must be over 40 per cent of the wavelength. The second case is when x becomes very small; the result is that $\tanh x = x$. Thus equation 8.19 reduces to

$$c = U \pm (gh)^{1/2}. \tag{8.21}$$

When h is sufficiently small (about $\frac{1}{40}$) compared with λ, equation 8.21 gives the velocity of 'long' waves in shallow flows.

Inertia waves

In contrast to sound and gravity waves, inertia waves are due to the rotation of the earth. The waves are so called because the Coriolis force is due to the inertia of moving masses on the earth. By eliminating the effects of gravity and density variations, and by considering a layer of fluid of depth h, perturbation equations for u, v and w together with a two-dimensional continuity equation may be derived. Their solution gives waves with speeds

$$c = \frac{\omega \sin \phi \lambda}{\pi \left(1 + \frac{4h^2}{\lambda^2} \right)^{1/2}},$$

where ϕ is latitude. Clearly, when $\omega = 0$ (i.e. no earth rotation), $c = 0$ yet a stationary wave pattern of streamlines could exist.

Rossby waves

As the large-scale flow in the extra-tropical atmosphere takes on a wave form (see figure 8.1) it is not surprising that the search of a theoretical explanation in terms of waves exercised the meteorologists of the inter-war period. A major breakthrough was provided by C-G. Rossby (1939) who, as Harwood (1981) has outlined, showed that the principle of conservation of absolute vorticity largely accounted for the wave-like configuration of flow. We are now in a position to re-appraise the Rossby regime purely in terms of waves. In the previous section we showed that the earth's rotation may generate inertia waves. Rossby went on to show that the latitudinal variation of the vertical component of the earth's vorticity (or Coriolis parameter, f) is of particular significance to the large-scale waves in the extra-tropical atmosphere. These waves are horizontal-transverse, which, although of great importance from the viewpoint of meteorology are virtually unknown in fluid dynamics outside the subject. Some of their characteristics can again be analysed by perturbation methods.

We start once more with the equations of motion as typified by equation 8.2. To suppress vertical-transverse waves (gravity waves), we require that the air particle orbits lie in horizontal planes, and to exclude compression or sound waves, we regard the atmosphere as an incompressible and horizontally homogeneous fluid. Because of these constraints, $w = 0$ and $\partial p / \partial x = \partial p / \partial y = 0$. In turn this means that equation 8.2 reduces to

$$\frac{\partial u}{\partial t} + u\frac{\partial u}{\partial x} + v\frac{\partial u}{\partial y} - fv + \frac{\partial}{\partial x}\left(\frac{p}{\rho}\right) = 0. \tag{8.22}$$

We can apply exactly the same kind of arguments to the velocity (v) in the meridional direction (y) and the resultant equation of motion for that component becomes

$$\frac{\partial v}{\partial t} + u\frac{\partial v}{\partial x} + v\frac{\partial v}{\partial y} + fu + \frac{\partial}{\partial y}\left(\frac{p}{\rho}\right) = 0. \tag{8.23}$$

The equation of continuity (Panofsky, 1981, equation 2.2), together with the condition for incompressibility, means that

$$\frac{\partial u}{\partial x} + \frac{\partial v}{\partial y} = 0. \tag{8.24}$$

Differentiating equation 8.22 with respect to y and equation 8.23 with respect to x and subtracting the first equation from the second we get

$$\frac{\partial}{\partial t}\left(\frac{\partial v}{\partial x} - \frac{\partial u}{\partial y}\right) + u\frac{\partial}{\partial x}\left(\frac{\partial v}{\partial x} - \frac{\partial u}{\partial y}\right) + v\frac{\partial}{\partial y}\left(\frac{\partial v}{\partial x} - \frac{\partial u}{\partial y}\right) + u\frac{\partial f}{\partial x} + v\frac{\partial f}{\partial y}$$

$$+ \left(f + \frac{\partial v}{\partial x} - \frac{\partial u}{\partial y}\right)\left(\frac{\partial u}{\partial x} + \frac{\partial v}{\partial y}\right) = 0. \tag{8.25}$$

This rather fearsome equation can fortunately be simplified and re-written in a less off-putting way. First, from equation 8.24 we know immediately that the last term in equation 8.25 vanishes. Secondly, we know from Harwood (1981) that the expression $\left(\dfrac{\partial v}{\partial x} - \dfrac{\partial u}{\partial y}\right)$ is the vertical component of the vorticity of airflow, denoted by Harwood as ζ_r. Thirdly, we know that the Coriolis parameter ($f = 2\omega \sin \phi$) changes with neither time (the earth spins at a constant rate) nor longitude. This means that $\partial f/\partial t = 0$ and $\partial f/\partial x = 0$ and consequently the addition of these terms to an equation will not alter its equality. For example, the first term on the left-hand side of equation 8.25 may be initially re-written as $\partial/\partial t(\zeta_r)$. It may also be written as $\partial/\partial t(\zeta_r + f)$ because we have just shown that $\partial f/\partial t = 0$. Using this device we may re-write the whole of equation 8.25 as

$$\frac{\partial}{\partial t}(\zeta_r + f) + u\frac{\partial}{\partial x}(\zeta_r + f) + v\frac{\partial}{\partial y}(\zeta_r + f) = 0. \tag{8.26}$$

Since we specified that $w = 0$, equation 8.26 further reduces to

$$\frac{d}{dt}(\zeta_r + f) = 0. \tag{8.27}$$

Equation 8.27 is a particular form of equation 4.5 in Harwood (1981).

To establish the phase speed of the horizontal-transverse waves we now constrain them to travel in the x direction by letting u and v be independent of the y (meridional) co-ordinate. With these restraints, ζ_r reduces to $\partial v/\partial x$. Remembering that f is independent of x and t, equation 8.26 becomes

$$\frac{\partial^2 v}{\partial x \partial t} + u\frac{\partial^2 v}{\partial x^2} + \beta v = 0, \tag{8.28}$$

where $\beta = \partial f/\partial y$, known as the Rossby parameter and given by $2\omega \cos \phi/a$ where a is the radius of the earth. To simplify the problem the latitudinal variation of β is neglected. As we have taken v to be independent of y, equation 8.24 reduces to

$$\frac{\partial u}{\partial x} = 0,$$

The problem is further simplified by assuming that the zonal current has a speed U, independent of x, y and t. With these simplifications, equation 8.28 has wave solutions of the form

$$v = A \cos \frac{2\pi}{\lambda}(x - ct),$$

where A is the amplitude of the meridional wind component. Insertion of this solution into equation 8.28 gives a wave speed

$$c = U - \frac{\beta\lambda^2}{4\pi^2}. \tag{8.29}$$

This equation is Rossby's wave equation, giving the speed of transverse waves in the predominantly horizontal motion of the atmosphere. Equation 8.29 tells us that, in contrast to sound and gravity waves, Rossby waves are always propagated 'upstream', or westward relative to the medium and travel at speeds that depend on their wavelengths. They also travel relatively slowly; for wavelengths of 5000 km their speed relative to the medium is of the order of 4 m s^{-1}. Note that the wave equation cannot account for waves which travel faster than the wind. Equation 8.29 does however show that stationary waves can exist when the wavelength is long. The stationary wavelength is given by

$$\lambda_s = 2\pi \left(\frac{U}{\beta}\right)^{1/2} .$$

The larger the mean wind speed U, the longer the wavelength of these stationary waves. At a latitude of $45°$, $\beta \simeq 16 \times 10^{-12}$ m^{-1} s^{-1} and for a typical winter value of U at 500 mb of 20 m s^{-1} the stationary wavelength is about 700 km. A typical summer value of 10 m s^{-1} gives a wavelength of about 5000 km. These surprisingly good results emerged despite three strong constraints on Rossby's analysis. He assumed that first the flow was horizontal and non-divergent; secondly, the atmosphere was autobarotropic so that the basic current U was constant with height (in addition, U was assumed to be uniform meridionally); and thirdly, the perturbation components u, v and p were independent of y, or that they had infinite lateral extent.

Baroclinic waves

Observations show that wind speeds in the extra-tropics increase with height. This occurs primarily because the extra-tropical atmosphere is *baroclinic* i.e. surfaces of constant pressure and constant density intersect each other. Whereas Rossby's theory did not encompass vertical wind shear, classical papers by Charney (1947) and Eady (1949) analysed the extra-tropical wave structure in a baroclinic atmosphere. It is inappropriate to go through lengthy and, in places, complicated mathematics used in their analyses but we can give a brief outline of the main results, arrived at, we may add, by the application of perturbation techniques to a more complete form of the vorticity equation than equation 8.26. Complex wave speeds were derived, giving information on phase speed and stability. These parameters were found to be related to the wavelength and the degree of vertical wind shear in the flow. Figure 8.5 summarizes the main results. For wavelengths less than about 3000 km the waves are stable for all values of wind shear. Even when the wavelengths are longer they are stable for shear values of about 1 m s^{-1} km^{-1}. However, where the shear is greater than about 1 m s^{-1} km^{-1}, there exists a band of wavelengths for which instability will set in. The main result of this analysis is than an intermediate band of wavelengths will be unstable for sufficiently large values of the wind shear. The horizontal line on figure 8.5 represents a typical value of vertical shear (2 m s^{-1} km^{-1}) and we see that it intersects the unstable region. This means that the extra-tropical westerlies

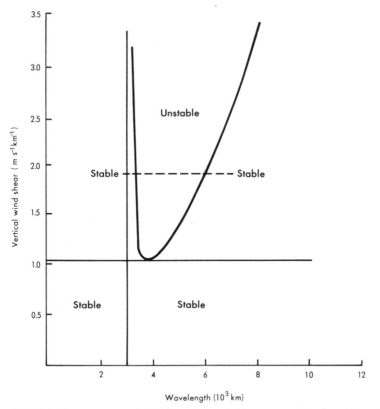

Fig. 8.5 Occurrence of stable and unstable waves as a function of wavelength and vertical wind shear (after Wiin-Nielsen, 1973).

are *always unstable* with respect to disturbances of the kind considered here. This is a very fundamental result that affects all weather and climate in the middle and high latitudes. Further analysis shows that the unstable waves will increase two- to three-fold in amplitude in about two days, a result in accord with observation of the real atmosphere. It is interesting to note that these profound insights gained from analytical techniques appeared at just the same time as the first application of numerical techniques to atmospheric analysis.

Conclusion

In this chapter, we have outlined some of the theoretical insights into atmospheric behaviour gained from the use of perturbation methods. We have seen that careful manipulation of the fundamental meteorological equations reveals the significant effects of, among other things, gravity, the rotation of the earth, the latitudinal variation of that rotation, and vertical wind shear upon atmospheric motion. The elementary perturbation method concentrates on wave characteristics of fluid flow, but we have seen that such results have direct relevance to an understanding of atmospheric behaviour. Further insights have emerged from the application of numerical techniques, a topic covered later in this book.

References

Atkinson, B. W. (1981) 'Weather, meteorology, physics, mathematics', this volume, 1–7.

Charney, J. G. (1947) 'The dynamics of long waves in a baroclinic westerly current', *J. Met.*, 4, 135–62.

Eady, E. T. (1949) 'Long waves and cyclone waves', *Tellus*, 1, 35–53.

Gadd, A. J. (1981) 'Numerical modelling of the atmosphere', this volume, 194–204.

Harwood, R. S. (1981) 'Atmospheric vorticity and divergence', this volume, 33–54.

Panofsky, H. A. (1981) 'Atmospheric hydrodynamics', this volume, 8–20.

Richards, J. P. G. and Williams, R. P. (1972) *Waves*, Harmondsworth, Penguin Books.

Rossby, C-G. and collaborators (1939) 'Relation between variations in the intensity of the zonal circulation of the atmosphere and the displacement of the semi-permanent centres of action', *J. Mar. Res.*, 2, 38–55.

Thompson, P. D. (1961) *Numerical Weather Analysis and Prediction*, London, Macmillan.

Wiin-Nielsen, A. (ed.) (1973) *Compendium of Meteorology*, Vol. 1, Geneva, World Meteorological Organisation.

Additional reading

Haltiner, G. J. and Martin, F. L. (1957) *Dynamical and Physical Meteorology*, New York, McGraw-Hill.

Haurwitz, B. (1941) *Dynamic Meteorology*, New York, McGraw-Hill.

Panofsky, H. A. (1968) *Introduction to Dynamic Meteorology*, Pennsylvania, Pennsylvania State University.

9
Dynamical meteorology: some milestones

B. W. ATKINSON
Queen Mary College,
University of London

Up to this point in the book, we have attempted not only to outline some physical theories and practical methods that are useful to meteorology, but also to show why they are of any use at all to the study of the atmosphere (Atkinson, 1981a). On the assumption that words such as 'hydrostatic', 'adiabatic', 'vorticity' and 'divergence' are no longer completely incomprehensible, it is perhaps worth while 'pausing for thought' and taking a look at some of the milestones of dynamical meteorology. In so doing, we shall concentrate upon the theoretical, as opposed to observational, developments over the century 1850–1950. Emphasis will lay on the general application of physics to the dynamics of the atmosphere, in particular its large-scale motion, and on the major analytical breakthroughs. It was, of course, only after 1950 that numerical models blossomed in meteorology and these are dealt with by Gadd (1981). In tracing the main avenues of progress the approach is chronological, concentrates on the contributions of selected notable scientists and suggests sources that the interested reader may wish to investigate further.

A century of analysis

As hinted above the century 1850–1950 saw the birth and coming of age of theoretical meteorology. According to Reed (1977, p. 391), 'Ferrel's work on the winds and the general circulation, done in the 1850s, can be said to mark the beginning of theoretical or dynamical meteorology'. The next century saw major developments by, among others, Peslin, Reye, Helmholtz, Guldberg and Mohn, V. Bjerknes, Margules, Richardson, Jeffreys, Rossby, Sutcliffe, culminating in the baroclinic wave theories of Charney and Eady in the late 1940s. Platzman (1968) highlights the end of an era (albeit a little early as he was discussing Rossby's work in the late 1930s) by noting that, as a result of Rossby's effort, '. . . it was now only one step to the barotropic model first used in computerized numerical weather prediction – a step that, although short, crossed the threshold into the realm of non-linear equations and thus severed the final link with the past' (Platzman, 1968 p. 228).

116

Useful secondary sources

Before concerning ourselves with the major developments of dynamical meteorology up to 1950, it is perhaps worth while emphasizing some particularly useful secondary sources. Meteorology is quite well served from a historical viewpoint. Frisinger's (1977) *History of Meteorology to 1800* provides a useful preface to Kutzbach's (1979) history of meteorological thought in the nineteenth century – a most valuable and readable volume. Pages of the *Bulletin of the American Meteorological Society* and *Weather* are occasionally devoted to the history of the subject. Worthy of particular mention is Platzman's (1967) masterly review of Richardson's (1965) classic entitled *Weather Prediction by Numerical Process*, originally published in 1922, but reprinted in 1965. Platzman (1968) has also applied his historical talents to the Rossby wave. Reed's (1977) Bjerknes Memorial Lecture is a stimulating and brisk treatment of the development of modern weather prediction. Readers may also find of interest the works by Brunt (1951), McIntyre (1972), Sawyer (1978) and the special issue of *Geofysiske Publikasjoner*, 24, 1962, in memory of V. Bjerknes. Finally, no bibliography of the history of meteorology, however brief, would be complete without mention of N. Shaw's *Manual of Meteorology*, published by Cambridge University Press in the late 1920s and early 1930s.

Early years

The roots of nineteenth century dynamical meteorology lay, perhaps as one would expect, in the works of eighteenth century physical scientists and mathematicians. Following in the footsteps of Halley and Hadley, D'Alembert 'was the first to attempt to express mathematically the motions of the atmosphere. His work was one of the first manifestations of a trend in meteorology which became more and more prevalent in the nineteenth century' (Frisinger, 1977, p. 130). Frisinger further notes that the basic physical and mathematical tools needed to study the atmosphere were well developed by 1800. These tools are, in essence, the equations of hydrodynamics and thermodynamics outlined by Panofsky (1981a, b) in earlier chapters. The reshaping of these tools to allow the application of Newtonian mechanics, in the guise of partial differential equations, to the atmosphere here owed much to the efforts of L. Euler (1707–1783). Euler developed the equations of motion in the form which is now familiar to physical scientists. As Frisinger (1977, p. 139) notes: 'Euler's development of the equations of motion was a great step in modern fluid dynamics. In this development he made the transition from parcels to points. No longer was it necessary to follow each parcel to calculate the forces upon it and thus deduce its accelerations. Instead, the calculation could be performed in a fixed co-ordinate system where rates of change at a point could be determined'. The implications of this breakthrough are exploited in the chapter by Gadd (1981). Frisinger further points out that the one ingredient missing from the set of equations governing the atmosphere, as known at the end of the eighteenth century, was the first law of thermodynamics, which was not formulated until well into the nineteenth century. To quote Frisinger (1977, p. 140), 'Thus, although scientists were ready by the beginning of the nineteenth century to study the mechanical aspects of atmospheric flow, they were not quite ready to study its thermodynamic properties. The study of these thermodynamic properties, along with the study of the atmospheric applications of the mechanical equations of motion, was the main task of nineteenth-century meteorology'.

Nineteenth century

The details of these nineteenth-century developments are elaborated by Kutzbach (1979). The studies of atmospheric thermodynamics were primarily concerned with the energetics of cyclones. Largely due to the American, Espy, the thermal hypothesis of cyclone formation was accepted for at least fifty years in the nineteenth century. According to this hypothesis, the initial diminution of pressure in the central part of a cyclone was attributed to heating caused by the release of latent heat during condensation of water vapour in ascending currents of warm moist air. It was only around the turn of the nineteenth century that this hypothesis was found to be seriously deficient. Assessments of the thermal hypothesis from an energetic point of view shifted emphasis from the latent heat release, as a primary source of kinetic energy in cyclones, to the advection of warm and cold air, as the major source of localized heating. Max Margules confirmed this view in 1903 by quantitative calculations based on simple models of initially unstable mass distributions of warm and cold air. He showed that adiabatic redistribution of the air masses, associated with rising of warm air and sinking of cold air to a state of thermal stability, would lead to the lowering of the centre of gravity of the whole system and conversion of potential into kinetic energy. These ideas are further elaborated in the chapter by White (1981).

Although the study of atmospheric thermodynamics proceeded apace, hydrodynamics were not ignored. Indeed their application to the atmosphere first appeared in the mid-nineteenth century. Building on Laplace's general equations of motion of a particle relative to a rotating earth, and being aware of the Coriolis effect, the American, William Ferrel, presented the first mathematical formulation of atmospheric motions on a rotating earth in 1859. Later, in 1878, he derived the important concept of the thermal wind which shows the relationship between the horizontal temperature gradient and the vertical shear of the wind.

Bjerknes' circulation theory

Around the turn of the century, V. Bjerknes, globally recognized as a father-figure of modern meteorology, applied to the subject the theorem of velocity circulation which had been developed earlier for homogeneous and incompressible fluids. In contrast to earlier investigators, who generally followed classical hydrodynamics and treated air density as a function of air pressure in studies of atmospheric dynamics, Bjerknes realised that atmospheric density is not constant along pressure surfaces. Such a situation where density and pressure surfaces intersect is described as *baroclinic*. Bjerknes' circulation theorem is applicable to such situations and allows the quantitative evaluation, from observational data, of the thermal energy available for circulation. This relatively simple, yet powerful, equation has had a profound effect on meteorology and is worthy of a brief outline here. It is impossible to give full details of the nature of circulation *per se* and the circulation theorem in this brief article. The interested reader is referred to a basic text such as that by Hess (1959). Nevertheless, as Harwood (1981) has already explained the meaning of vorticity, it should not be too difficult to conceive *circulation* as a measure of the extent to which a fluid exhibits rotary motion — positive for cyclonic circulation and negative for anticyclonic circulation. As perhaps one would expect, there is a relationship between vorticity and circulation: for a small element of area the vorticity of the fluid in the area is equal to the circulation around the perimeter per unit area. Put slightly differently, circulation is an areal measure of the rotational tendency of a fluid and vorticity is a point measure of that same tendency. In his circulation theorem, Bjerknes showed that circulation would change with time (positively or negatively) because of three effects: the so-called solenoid effect due to the intersection of density and pressure surfaces; the Coriolis effect due to the rotation of the earth; and a general effect of forces other than pressure-gradient, Coriolis and gravity forces. The solenoid effect is given by the term $-\oint\dfrac{dp}{\rho}$, where p is pressure, ρ is density and where the integration is carried out around a closed path. The illustration used here is taken from Hess (1959). Figure 9.1 shows a situation where the lines of equal density and equal pressure intersect. Such a situation could occur across a coastline in the boundary layer or a front in the free atmosphere. Evaluation of $-\oint\dfrac{dp}{\rho}$, or at least some indication of its magnitude and sign would tell us whether circulation exists and in what direction the air is moving. Following Hess, we can investigate the solenoid effect around the rectangle 1, 2, 3, 4. In going from pt 1 to pt 2 and from pt 3 to pt 4, pressure does not change, so $dp = 0$ and consequently, there is no contribution to circulation change from these quarters. On the leg from pt 2 to pt 3, dp is negative so the contribution to the integral must be positive (note the negative sign) while, on the leg from pt 4 to pt 1, dp is positive, so that the contribution to the value of the integral is negative. But the mean density along 2 to 3 is less than that along 4 to 1. Because the density appears in the denominator of the integrand the positive contribution (2 to 3) overcomes the negative contribution (4 to 1) and the entire term makes a positive contribution to dc/dt (the change in circulation, c). Positive circulation is anticlockwise so the lines of constant density will turn in that sense in an attempt to become parallel to the isobars. In doing this, dense air sinks and less dense air rises. Such a movement would lower the centre of gravity of the whole body of air, thus reducing the potential energy of the system, and, by the conservation of energy, increasing the kinetic energy. A pressure and density distribution such as illustrated in figure 9.1 could well occur if point 2 were on land and point 1 at sea. We have seen that and the resultant change in circulation would encourage air flow in the direction 1, 2, 3,

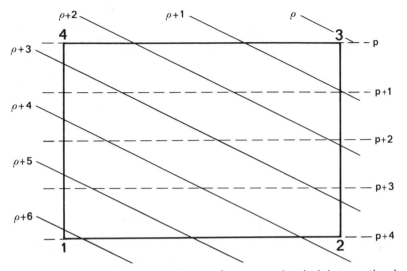

Fig. 9.1 Isobars and isosteres (lines of constant density) intersecting in a baro-clinic fluid (after Hess, 1959).

4: of that, the part flowing from pt 1 to pt 2 would be the familiar sea breeze. Before leaving this topic it is important to note that the theorem deals only with temporal *changes* of circulation, not with the circulation itself. Thus, the contribution from the solenoid term may be positive yet the circulation be negative. A circulation that actually moves in the same sense as the direction in which the solenoid effect drives it (as in the case of the sea breeze circulation) is known as a *direct circulation*. A circulation that is actually moving in the opposite sense from the direction in which the solenoid effect impels it (such as happens in the mid-latitude Ferrel cell of the general circulation) is called an *indirect circulation*.

The importance of the circulation theorem was not doubted by T. Bergeron a young collaborator of V. Bjerknes. In later years, Bergeron (1962, p. 11–12) wrote: 'The circulation theorems threw a bridge between classical hydrodynamics . . . and classical thermodynamics . . . . The foundations were laid for a *physical hydrodynamics* (his italics), by which the step could be taken from the writing desk of the theoretician, and from the laboratory of the experimentalist, out into reality: to the atmosphere and oceans of our rotating planet . . . . The baroclinic developments, i.e. the circulation accelerations, are vital phenomena, nowadays (i.e. 1962) being taken up again for study by the foremost theoreticians. These processes will, of course, eventually form the main concern of dynamic (sic) meteorology and weather forecasting'.

Richardson

We noted earlier in this chapter that our primary interest would be in the results of analytical, as opposed to numerical methods. Nevertheless, one cannot even briefly scan the history of meteorological theory without noting the work of L. F. Richardson. In 1922, having worked on the subject throughout the 1914—18 war, Richardson published his book entitled *Weather Prediction by Numerical Process.* In this volume he 'solved' the meteorological equations by numerical as opposed to analytical methods. In so doing, he was thirty years ahead of his time. His solution (in the form of a surface pressure change of 145 mb in six hours) was of course seriously wrong, but the overall strategy forms the basis of to-day's numerical forecasting (see chapter by Gadd, 1981). Richardson's failure resulted partly from inadequacies in upper wind data at the time of his experiment and partly from the impossibility of using observed winds (even if available) to calculate pressure change from the pressure-tendency equation. It is similarly impossible to do justice to Richardson in this article and the interested reader is strongly urged to consult the superb review by Platzman (1967). But two further points are worth making. First, Platzman (1967, p. 530) expresses the view 'that a major defect in Richardson's approach to weather prediction is his disregard of perturbation theory as a means of clarifying the problems of dynamic (sic) meteorology through analysis of atmospheric wave motion (see Atkinson, 1981b). Although Richardson's book came slightly before the great developments which took place along these lines at the hands of V. Bjerknes and his Norwegian school, nevertheless it is strange that little of perturbation theory is visible in the book . . . . If this is a characteristic feature of his scientific style, perhaps some of the major defects in his prediction scheme were inevitable. However, without a conditioning against perturbation theory the whole conception of the book might not have germinated and some of his major contributions to meteorology might never have emerged!' Secondly, and somewhat related, Jeffreys (1922) commented that 'The method . . . is necessarily laborious . . . and probably could not be worked with sufficient speed to make it a practical method of forecasting; but when forecasters have acquired experience in its use, they will probably find that a sufficient number of quantities allowed for are comparatively small to make it possible to expedite the calculation considerably without great sacrifice of accuracy'. Although Jeffreys was not a forecaster he did soon after investigate the relative sizes of the terms in the equations of motion with most interesting results — now to be considered.

Jeffreys and the dynamics of winds

The first chapter in this series (Atkinson, 1981a) revealed that Newton's second law can be applied to the atmosphere. Panofsky (1981) went on to show that the acceleration of air is due to the sum of several forces, and presented a simplified equation to describe the situation. It is now appropriate to consider how that simplification came about and, further, to contemplate some implications of the equations as seen by Jeffreys. To do this, we must start by considering rather lengthy forms of the equations of motion (known as the primitive equations), one showing the forces and accelerations of air in a west—east (or vice versa) direction (known as zonal direction) and one showing the situation for vertically moving air. North–south motions are governed by an equation similar in form to that governing west–east motion.

The equation for zonal flow is as follows:

$$\frac{du}{dt} - \frac{uv}{a}\tan\phi + \frac{uw}{a} = -\frac{1}{\rho}\frac{\partial p}{\partial x} + 2\Omega(v\sin\phi - w\cos\phi) + Fr. \qquad (9.1)$$

(Accelerations)   (Forces)

The equation for vertical motion is

$$\frac{dw}{dt} - \frac{(u^2 + v^2)}{a} = -\frac{1}{\rho}\frac{\partial p}{\partial z} - g + 2\Omega u\cos\phi + Fr.$$

(Accelerations)   (Forces)   (9.2)

In these equations, u is the zonal component of air velocity, v is the meridional (north–south and vice versa) component of air velocity, w is the vertical component of the air velocity, t is time, ϕ is latitude, a is the radius of the earth, ρ is the air density, p is pressure, x is the zonal co-ordinate, Ω is the angular velocity of the earth $(7.29 \times 10^{-5}\ \text{s}^{-1})$, g is gravity and Fr is the frictional force. On the left-hand side of equations 9.1 and 9.2, the first terms show the acceleration of the air relative to the earth and the remaining terms show the acceleration due to the curvature (not the rotation) of the earth's surface. On the right-hand side of equation 9.1 the terms are the pressure-gradient force, the Coriolis force (due to the earth's rotation) and the frictional force respectively: equation 9.2 has the appropriate components of the same forces together with the effect of gravity.

We are now in a position to estimate roughly the relative size of these terms, knowing, as we do, typical values of u, v $(10\ \text{m s}^{-1})$, w $(5\ \text{cm s}^{-1})$, p (giving δp of about 1 mb/100 km in the horizontal), ρ (depends on altitude but near the ground about 1 kg m^{-3}) and the actual values of a (6371 km) and Ω. If we take $\phi = 45°$ and work in the c.g.s. system the following results:

$$\frac{du}{dt} \qquad -\frac{uv}{a}\tan\phi \qquad +\frac{uw}{a} \qquad = -\frac{1}{\rho}\frac{\partial p}{\partial x} \qquad +2\Omega \quad (v\sin\phi - w\cos\phi) + Fr\ .$$

$$\frac{10}{10^3} \qquad \frac{10^6}{6\times 10^{-8}} \qquad \frac{5\times 10^3}{6\times 10^8} \quad \frac{10^3\times 10^4}{10^8} \qquad 10^{-4} \quad (10^3 \qquad\qquad 5) \qquad ?\quad \text{cm s}^{-2}$$

$$10^{-2} \quad 1.6\times 10^{-3} \quad 10^{-5} \qquad\quad 10^{-1} \qquad\qquad 10^{-1} \qquad\qquad 5\times 10^{-4} \quad ?\quad \text{cm s}^{-2}.$$

Terms in the middle line are derived by inserting the typical values noted above and approximating to get orders of magnitude. Thus, in the first term on the left-hand side, observation tells us that wind speeds vary little over a period of about quarter of an hour. In the second term on the right-hand side, we know that $\Omega = 7.29 \times 10^{-5}\ \text{s}^{-1}$: consequently $2\Omega \simeq 14 \times 10^{-5}\ \text{s}^{-1}$ which is $1.4 \times 10^{-4}\ \text{s}^{-1}$, i.e. of the order of $10^{-4}\ \text{s}^{-1}$. The magnitudes of the frictional terms are even less well known than some of the others (hence the question mark) but are probably of order 10^{-5} or less. If we now concentrate on the third line we see that some terms are far bigger than others (ten times bigger/smaller is called 'an order of magnitude' bigger/smaller). On the left-hand side we can justifiably retain only the first term, whilst on the right-hand side those terms of order 10^{-1} are retained. Hence we have justified the simplified form used in the earlier chapter by Panofsky. It is the pressure-gradient force, the part of the Coriolis force due to horizontal flow, and possible frictional force that are the primary causes of the major acceleration term, i.e. the air flow relative to the earth.

If we apply the same arguments to the equation of vertical motion the following ensues:

$$\frac{dw}{dt} - \frac{(u^2 + v^2)}{a} = -\frac{1}{\rho}\frac{\partial p}{\partial z} \quad -g \quad +2\Omega u \cos\phi \quad +Fr.$$

$$\frac{10^{-1}}{10^3} \quad \frac{2 \times 10^6}{6 \times 10^8} \quad \frac{10^3 \times 5 \times 10^5}{6 \times 10^5} \quad 10^3 \quad 10^{-4} \times 10^3 \quad ? \quad \text{cm s}^{-2}$$

$$10^{-4} \quad 3 \times 10^{-3} \quad 10^3 \quad 10^3 \quad 10^{-1} \quad ? \quad \text{cm s}^{-2}.$$

In this case only the terms of order 10^3 need be retained and this gives us the hydrostatic equation as outlined by Panofsky, viz.

$$-\frac{1}{\rho}\frac{\partial p}{\partial z} = g \quad \text{or} \quad \frac{\partial p}{\partial z} = -\rho g.$$

In the first of his two classical meteorological papers in the 1920s, Jeffreys (1922) employed scale analysis such as outlined above in his classification of winds according to their dynamics. He used the simplified equations but retained the frictional forces. He acknowledged that 'the frictional terms are not certainly known' but, despite this deficiency, assured the reader that 'the accuracy of the equations nevertheless much exceeds that required' (Jeffreys, 1922, p. 31). On the basis of the equations, which for simplicity, may be written here in the form Acceleration = press-grad force + Coriolis force + friction force, Jeffreys classified winds into three types assuming that all types were initiated by the pressure-gradient force. First, he identified as Eulerian winds those occurring when the Coriolis and frictional terms are small in comparison with the acceleration term (du/dt). Secondly, if the Coriolis terms are far larger than both the acceleration and frictional terms, the resultant wind is the familiar 'geostrophic' wind, as outlined by Panofsky, where the pressure-gradient and Coriolis forces balance. Thirdly, if the frictional terms are larger than the Coriolis and accelerational terms, the resultant wind was called 'antitriptic' by Jeffreys. Note that, in all three types, Jeffreys considered the pressure-gradient force to be important. In fact, it is possible to have movement relative to the earth in the absence of a pressure-gradient force, such as the circles of inertia exhibited by the GHOST balloons. Jeffreys was aware that he had only partially solved the problem: 'Besides these main divisions there will be many cases where the number of terms in the equation of motion comparable with the pressure term is two or three. In such cases, the motion is composite in character and intermediate between the three main types. The elucidation of the simpler cases must, however, afford some assistance towards the understanding of the more complex ones' (Jeffreys, 1922, p. 34). Jeffreys went on to use his classification to elucidate the dynamics of airflows of different horizontal scale, ranging from 'the general circulation and its seasonal variation' (Jeffreys, 1922, p. 34) to small-scale phenomena such as sea and mountain breezes. He considered tropical cyclones to be examples of Eulerian winds, that synoptic-scale flows tended to be geostrophic and that the small-scale circulations such as sea breezes exemplified antitriptic winds. Whilst some of these results would be queried to-day, the clarity of thought and the genetic, dynamical approach to classifying winds means that Jeffreys' paper is still well worth reading to-day.

Four years later, Jeffreys (1926) produced another major paper on the dynamics of geostrophic winds. Sawyer (1978, p. 253) admirably summarizes the major contribution of the article. Jeffreys 'showed that the exchange of air between the tropics and polar latitudes which is needed to carry the excess heat from the equator towards the poles cannot be achieved by meridional circulations alone – that is, not by air moving generally polewards at one level and back towards the equator at another. He showed that the exchange has to be carried out by deep southerly air currents in one longitude and return northerly currents in other longitudes. This is a consequence of the balance of angular momentum about the earth's axis within latitude bands of the atmosphere. Although the ideas in Jeffreys' paper were not followed up for many years afterwards, his ideas of the essential role played by large-scale eddies (such as depressions and anticyclones) now form an integral part of our understanding of the general circulation of the atmosphere and the basis for the quantitative study of the earth's climate'.

Cyclones and pressure changes

Perhaps the neglect of Jeffreys' contributions was due to the flush of enthusiasm for the then recent developments from V. Bjerknes' Norwegian school of meteorology. The concepts of polar front and frontal cyclone which had been presented to the world at the beginning of the 1920s tended to over-shadow larger-scale, perhaps even more fundamental dynamical studies such as those by Jeffreys. Indeed the 'cyclone problem', with its many facets, had a significant effect on the direction taken by research in dynamical meteorology throughout the inter-war years. Mid-way through that period, Brunt (1930) recognized two main theories of the origin of cyclones: first, the 'local heating' theory in which instability due to local heating from the underlying topography led to uplift of the low-level converging air to higher levels where it was somehow whisked away out of the cyclone; and secondly, the familiar 'polar-front' theory, within which the primary cause of uplift was more dynamical (but not then fully specified) than thermal. V. Bjerknes (1921) was the first to theorize on the newly found frontal cyclone, soon to be followed by Brunt (1924) himself. The facet of the problem which eventually became a major interest was the mechanism of pressure change, which, in the context of cyclones, meant answering the question of how is the low pressure in the centre of cyclones initiated and maintained.

For an introduction to the problem, we turn once more to Jeffreys (1919) who showed that, if we ignore the effects of latitude, geostrophic winds are non-divergent and thus cannot produce surface pressure changes. This result was forcefully championed by Sutcliffe (1938, p. 496), who noted that 'this fact . . . does not appear to have been explicitly stated in standard text-books'. It is amusing to note later in the discussion on Sutcliffe's paper that Brunt confessed to omitting it from his classical text-book *Physical and Dynamical Meteorology* 'through the accident of mislaying the sheet on which it had been written' (Brunt, 1938, p. 506). In fact, a decade earlier, Brunt and Douglas (1928) had shown that pressure changes, vertical motion and rain cannot be explained in terms of geostrophic winds alone. They showed that the deviation from geo-strophic is important and emphasized the relation of this ageostrophic component to the acceleration of the air. In addition, they showed that the ageostrophic wind could be estimated from the pressure field and its changes.

The same discussion also reveals an interesting example of how even eminent professional meteorologists occasionally do not fully comprehend new sugges-tions on their first hearing. To quote Brunt (1938, p. 506): 'Dr. Sutcliffe had pointed out that so long as the wind was geostrophic there could be no rise of pressure when cold air replaced warm air. It is, however, a matter of frequent occurrence than when warm air is replaced by cold air the pressure rises, and that when cold air is replaced by warm air, the pressure falls. That this is not always the case is at least in part due to the fact that the atmosphere is three-dimensional, and that the simple replacement we are considering is not the whole story'. To which Sutcliffe replied: 'Professor Brunt expressed an attitude to changes in surface pressure occurring with advection of air of different tempera-tures, which was precisely what Dr. Sutcliffe would like to see removed from meteorology . . . He saw no reason to attempt to explain why pressure changes by 'simple replacement' mentioned by Professor Brunt were not the 'whole story'; he would prefer to regard them as a fairy tale'. One year later Sutcliffe (1939) continued his crusade, praising Dines' (1913) work a quarter of a century earlier. In these early studies, Dines noted that if some mechanism could be intro-duced to account for divergence of air above a developing depression and converg-ence above an anticyclone within the upper troposphere, the observed motion systems would be fully explained. Sutcliffe (1939, p. 518) lamented that 'this pronouncement has been over-shadowed in recent work by the success of the frontal and air mass conceptions, which has been achieved in spite of the inade-quacy of dynamical theory . . . ' of the said frontal and air mass ideas. As a final comment before presenting his own theory, Sutcliffe (1939, p. 519) quotes Dines (1922): 'it is the mass of air which is important; its temperature is quite immater-ial'. Discussions such as the above give some small comfort to the non-theoretician who finds things difficult. A similar situation occurred eighteen years later in the context of numerical modelling when some of the intricacies of that then comparatively novel method had to clarified for Professor Brunt's distinguished successor (Discussion of Phillips, 1956).

Understanding of the mechanism of pressure change and thus, to a large measure, also of the development of cyclones and anticyclones emerged from the efforts of, among others, Bjerknes (1937), Bjerknes and Holmboe (1944), Sutcliffe (1938, 1939, 1947) and Houghton and Austin (1946). Using Sutcliffe's results as exemplification here, development involves low-level horizontal divergence (positive or negative) approximately balanced by divergence of opposite sign in the upper troposphere, the total divergence integrated vertically and represented by the rate of change of surface pressure being a relatively small residual (see figure 9.2). Consequently, a criterion for development of cyclones and anticyclones is that there should be a significant difference between the lower and upper fields of divergence.

Very closely related results emerged from the efforts of Bjerknes and Holmboe (1944). In particular, they analysed the relationship between the upper level Rossby waves, the distribution of horizontal divergence and cyclogenesis. They found that, in a wave-shaped westerly flow, with sufficiently strong westerlies, horizontal divergence and convergence occur in such a distribution as to cause eastward movement of the troughs and crests. Where the west winds increase with height (as is normal in mid-latitudes) cyclones form, develop thermal asymmetry and intensify. The basic reason for the pressure developments outlined by Bjerknes and Holmboe is essentially the same as that proposed by Sutcliffe. In Bjerknes and Holmboe's (1944, p. 1) own words: 'The main reason for the usual eastward drift of closed isobar patterns in temperate latitudes, lies . . . in the fact that the superimposed wave pattern in the upper layers produces overcompensating accumulation of air where the low levels show depletion, and depletion of air where the low levels show accumulation'.

The results of both Sutcliffe and Bjerknes and Holmboe married well with Rossby's findings on the planetary waves. It is convenient to recognize a difference in scale: on the one hand, the large Rossby waves, primarily a function of planetary vorticity, within which distinctive distributions of horizontal divergence and convergence occur; and on the other, the smaller cyclone waves which depend largely on these divergences for their existence. In the previous chapter (Atkinson, 1981b) we outlined the basics of Rossby waves (see also Bolin, 1959), together with a mention of the results of Charney (1947) and Eady (1949). To some extent, these latter two classical papers built on the ideas expressed by Sutcliffe

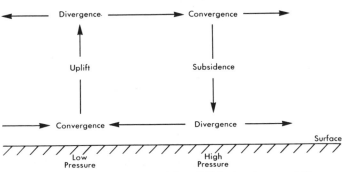

Fig. 9.2 Schematic diagram of Sutcliffe's development theory.

(1947) about the baroclinic instability in extra-tropical latitudes. Both Charney and Eady were concerned with assessing the stability of 'waves' (i.e. cyclones) within a westerly current. Figure 8.5 of Atkinson (1981b) shows their main results. Eady found that short enough waves are stable when the meridional slope of the disturbance stream surface is greater than that of the mean potential temperatures (see Green, 1981). On the other hand Charney found that, for sufficiently long waves, the latitudinal variation of the Coriolis parameter is dominant and for stability the disturbance stream surface must slope downward with increasing latitude. The two criteria together yield an intermediate wavelength of maximum instability, i.e. cyclones on that wavelength would develop. Here is a very fundamental relationship between unstable cyclone waves and the meridional potential temperature gradient (or looked at another way, the mean vertical shear of the wind).

Conclusion

In concluding this chapter, one is of course aware of its short-comings, including the necessarily restricted choice of topics for inclusion. Other authors would probably have included different material and given a different perspective. The main aim has been to provide for the uninitiated a glimpse of the thread of progress in the application of dynamical concepts to atmospheric behaviour up to 1950. Echoing Sutcliffe's comments as quoted earlier, much of this material never appears in those standard texts which are *not* intended for the specialist dynamical meteorologist. The last three decades have seen an explosion of meteorological theory. In the remainder of this book, it is impossible to do justice to it. But with our basic tools from earlier chapters and some perspective from this present one, we are in a position to tackle topics such as atmospheric spectra, turbulence, energetics, more thoughts on baroclinic instability and a brief look at numerical modelling – all important themes in present-day meteorology.

References

Atkinson, B. W. (1981a) 'Weather, meteorology, physics, mathematics', this volume, 1–7.

Atkinson, B. W. (1981b) 'Atmospheric waves', this volume, 100–115.

Bergeron, T. (1962) 'The Stokholm period' (part of an appreciation of V. Bjerknes), *Geofysiske Publikasjoner*, 24, 11–15.

Bjerknes, J. (1937) 'Theorie der aussertropischen Zyklonenbildung', *Met. Zeit.*, 54, 462–6.

Bjerknes, V. (1921) 'On the dynamics of the circular vortex with applications to the atmosphere and atmospheric vortex and wave motions', *Geofysiske Publikationer*, 2 (4).

Bjerknes, J. and Holmboe, J. (1944) 'On the theory of cyclones', *J. Met.*, 1, 1–22.

Bolin, B. (ed). (1959) *The Atmosphere and the Sea in Motion*, New York and Oxford, The Rockefeller Institute Press and Oxford University Press.

Brunt, D. (1924) 'The dynamics of cyclones and anticyclones regarded as atmospheric vortices', *Proc. R. Soc. Lond. Ser. A.*, 105, 70–80.

Brunt, D. (1930) 'The present position of theories of the origin of cyclonic depressions', *Quart. J. R. Met. Soc.*, 56, 345–50.

Brunt, D. (1938) Discussion in 'On development in the field of barometric pressure', R. C. Sutcliffe, *Quart. J. R. Met. Soc.*, 64, 505–6.

Brunt, D. (1951) 'A hundred years of meteorology', *Adv. Sci.*, 8 (30), 114–24.

Brunt, D. and Douglas, C. K. M. (1928) 'The modification of the strophic balance for changing pressure distribution, and its effect on rainfall', *Mem. R. Met. Soc.*, 3 (22).

Charney, J. G. (1947) 'The dynamics of long waves in a baroclinic westerly current', *J. Met.*, 4, 135–62.

Dines, W. H. (1913) 'Cyclones and anticyclones', *J. Scot. Met. Soc.*, 16, 304–12.

Dines, W. H. (1922) 'The cause of anticyclones', *Nature*, 110, 845.

Eady, E. T. (1949) 'Long waves and cyclone waves', *Tellus*, 1, 35–52.

Frisinger, H. H. (1977) *The History of Meteorology to 1800*, New York, Science History Publications and American Meteorological Society.

Gadd, A. J. (1981) 'Numerical modelling of the atmosphere', this volume, 194–204.

Green, J. S. A. (1981) 'Trough–ridge systems as slant-wise convection', this volume, 176–193.

Harwood, R. S. (1981) 'Atmospheric vorticity and divergence', this volume, 33–54.

Hess, S. L. (1959) *Introduction to Theoretical Meteorology*, New York, Henry Holt & Co., Inc.

Houghton, H. G. and Austin, J. M. (1946) 'A study of non-geostrophic flow with applications to the mechanism of pressure changes', *J. Met.*, 3, 57–77.

Jeffreys, H. (1919) 'On travelling atmospheric disturbances', *Phil Mag.*, 37, 1–8.

Jeffreys, H. (1922) 'On the dynamics of winds', *Quart. J. R. Met. Soc.*, 48, 29–46.

Jeffreys, H. (1926) 'On the dynamics of geostrophic winds', *Quart. J. R. Met. Soc.*, 52, 85–101.

Kutzbach, G. (1979) *The Thermal Theory of Cyclones: A history of meteorological thought in the nineteenth century*, New York, American Meteorological Society.

McIntyre, D. P. (ed.) (1972) *Meteorological Challenges – A history*, Ottawa, Information Canada.

Panofsky, H. A. (1981a) 'Atmospheric hydrodynamics', this volume, 8–20.

Panofsky, H. A. (1981b) 'Atmospheric thermodynamics', this volume, 21–32.

Phillips, N. A. (1956) 'The general circulation of the atmosphere: a numerical experiment', *Quart. J. R. Met. Soc.*, 82, 123–64. Discussion appears on 535–9.

Platzman, G. W. (1967) 'A retrospective view of Richardson's book on weather prediction', *Bull. Am. Met. Soc.*, 48, 514–51.

Platzman, G. W. (1968) 'The Rossby wave', *Quart. J. R. Met. Soc.*, 94, 225–48.

Reed, R. J. (1977) 'The development and status of modern weather prediction', *Bull. Am. Met. Soc.*, 58, 390–9.

Richardson, L. F. (1965) *Weather Prediction by Numerical Process*, New York, Dover Publications.

Sawyer, J. S. (1978) 'Some highlights of the Society's publications over 50 years', *Weather*, 33, 251–9.

Sutcliffe, R. C. (1938) 'On development in the field of barometric pressure', *Quart. J. R. Met. Soc.*, 64, 495–504.

Sutcliffe, R. C. (1939) 'Cyclonic and anticyclonic development', *Quart. J. R. Met. Soc.*, 65, 518–24.

Sutcliffe, R. C. (1947) 'A contribution to the problem of development', *Quart. J. R. Met. Soc.*, 73, 370–83.

White, A. A. (1981) 'Atmospheric energetics', this volume, 153–175.

10

The spectrum of atmospheric motions

E. R. REITER
Colorado State University

The word 'spectrum' immediately brings to mind the simple experiment in which a beam of light passes through a glass prism and spreads into the colours of the rainbow. In more precise physical terminology, white light, which consists of the contributions from many different wave-lengths, is separated into a neat array of these contributions by increasing wave-lengths from blue to red, called the 'spectrum'.

By analogy, we may regard the motions in our atmospheric environment to consist of a superposition of eddies of different sizes and intensity, starting from the Brownian motion of molecules, to the small wiggles in the rising plume of cigarette smoke, to the gusts whirling autumn leaves down the lane, to the magnificent billows and bubbles in a towering cumulus cloud, to the passage of cyclones and anticyclones over the British Isles, to the meandering of atmospheric flow around the hemisphere, even to the change in seasons reflected in the behaviour of this flow around the globe.

If we regarded each of these eddies of vastly different sizes separately, according to its cross-sectional dimension or 'wave-length', we would be in a position to decompose atmospheric motions into their 'spectrum'. Obviously, a glass prism held into the air would not accomplish the job. We have to look for different, in this instance mathematical, tools fondly referred to as 'spectrum analysis'.

Unfortunately, a common mortal, equipped with an adding machine and a slide rule, will hardly be in a position to analyse the spectrum of atmospheric motions. Not that the calculations are difficult, but they are highly involved and require the shuffling of large arrays of data. It takes electronic computers to handle the job adequately and, even there, the smaller ones may find their power overtaxed. This fact should not deter us, however, from taking a simplified look at the working and philosophy of spectrum analysis.

Suppose we had an anemometer with a short response time that gives us the fine-scale structure of all the gusts that pass over the measurement site. A record from this anemometer would probably resemble the trace shown in figure 10.1. Immediately, we recognize that the 'gustiness' over the station consists of small

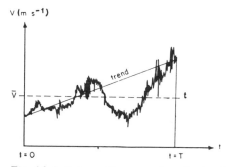

Fig. 10.1 Schematic example of a typical anenometer trace. Mean wind speed, $\bar{V}$, and linear trend are indicated by dashed and solid lines, respectively. Length of record = T.

but rapid wind variations, and of slower, more gradual shifts in wind speed (and direction) which, however, are of considerable amplitude. Finally, in the finite data sample shown here, the wind speed at time $t = 0$, $V_{t=0}$, differs markedly from the speed registered at the end of the record, $V_{t=T}$. In other words, during the time interval $t = T$, which characterizes the length of the record, the wind field that passed over the station underwent a certain 'trend' which may, or may not, have continued after the recorder stopped. The 'linear trend' between time $t = 0$ and $t = T$ may be computed simply as $\dfrac{V_{t=T} - V_{t=0}}{T} \times t$.

We now are faced with the task of estimating the contribution of each eddy size to the total 'gustiness' over the station within the record length T. Needless to say, this time length T may be chosen to our liking and needs. It could be one hour, one day, one year, several years. The choice will be dictated by the scales of atmospheric motions which we want to study in detail, by the money available for computer work, and by the number of data points which the computer is actually capable of handling.

In determining quantitatively the spectrum of light, we will have to measure the light intensity of each wave-length band across the rainbow, perhaps by moving a photoelectric cell along the 'spectrum' behind the glass prism. The intensity of an eddy in the spectrum of atmospheric motions is best measured by its kinetic energy, half the mass times the square of the velocity. If we take density as mass per unit volume, which is not supposed to vary strongly as the wind blows across our anemometer site, we may regard the square of the velocity (V^2), as our variable of prime interest that characterizes the intensity or energy of atmospheric eddies.

Let us return now to figure 10.1. By averaging the wind gusts over the time interval T, we obtain the *mean* wind speed, symbolized by an overbar, $\bar{V}$. All we have to do is to add together the individual wind speed readings, say every 1/10 s, and divide the sum thus obtained by the number of data points used in the summation. If we now compare *each* individual data point with the *mean* wind, $\bar{V}$,

131

we see that it differs by a certain departure symbolized by a prime, V'. Thus

$$V = \bar{V} + V' \tag{10.1}$$

where V' assumes positive and negative values. Since the record shown in figure 10.1 contains a strong trend of the wind speed with time, the values of V' are all negative at the beginning of the time interval, T, and they become increasingly positive towards the end of this time interval. This gives the impression of the values of V' being very large throughout the length of the record. In order not to give the false impression that the short gusts during the time interval, T, were extremely violent, we are wise to remove the linear trend from the record, thus considering the fluctuations V' shown in figure 10.2.

We can write the trend of the wind speed, $V^*(t)$, as a function of time

$$V^*(t) = V_{t=0} + \frac{V_{t=T} - V_{t=0}}{T} \times t. \tag{10.2}$$

The 'de-trended' wind fluctuations, shown in figure 10.2, may now be expressed as

$$V(t) - V^*(t). \tag{10.3}$$

It should be mentioned that, in actual practice, there are more sophisticated de-trending techniques available than the one outlined here. Their basic philosophy, however, remains the same.

Now we may proceed to regard the kinetic energy of the wind fluctuations, expressed by V'^2, as a function of eddy size. Instead of a glass prism, we use a mathematical or an electronic filter to accomplish the partition of energy into compartments arranged according to increasing eddy size.

We do this by sliding two narrow 'slots' along the wind record in figure 10.2, each revealing only one data point at a time. The distance Δt between these two 'slots' can be varied, say from 1/10 second, the time interval between the successive measurement points, all the way up to approximately $T/10$ or perhaps even $T/5$, where T is the length of the record. We now read the wind values that simultaneously appear in the two 'slots' as we move them along the whole record. The average difference between the wind speeds appearing in the two 'slots', as we move them along, gives us a measure of the energy contained in eddies that have a time duration of $2\Delta t$ as they pass over the measuring station. Assuming, with Sir Geoffrey Taylor, that these eddies float with the mean wind speed, $\bar{V}$, the average size of this particular set of eddies may be estimated as

$$\Delta x = \Delta t \bar{V}. \tag{10.4}$$

We can now vary the distance Δt between the 'slots' in figure 10.2 and repeat the whole operation, until we have covered all possible settings of Δt.

This sounds like a very cumbersome procedure. Fortunately, we do not have to carry out this operation manually by shifting 'slots' and reading off values from the wind record. We can leave this task to the computer which will come

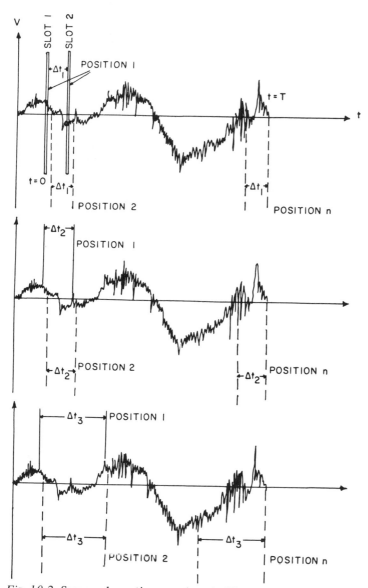

Fig. 10.2 Same schematic record as in Figure 10.1 except that mean wind, $\bar{V}$, and linear trend have been subtracted leaving only the wind fluctuations of interest. For an explanation of the meaning of slot positions see text.

up with the 'autocorrelation function', $R(\Delta t)$, that accomplishes the same thing.

$$R(\Delta t) = \frac{\overline{V'(t) \times V'(t + \Delta t)}}{\overline{[V'(t)]^2}}. \tag{10.5}$$

$V'(t)$ are the velocity fluctuations at time t about the de-trended mean value $\bar{V}$, shown in figure 10.2, and $V'(t + \Delta t)$ are the same fluctuations Δt seconds later. In other words, we *average* the product values $V'(t) \times V'(t + \Delta t)$ obtained at position 1, position 2, . . . position n, and divide this average value by the variance of the wind fluctuations within the data sample, $\overline{V'^2}$. The overbars in equation 10.5 signify averages of the respective values taken over the whole array of data. If everywhere in our data sample $V'(t) = V'(t + \Delta t)$, i.e. if the wind speed were constant during all the sampling time T, then $R = 1$. Also, for $\Delta t = 0$, R will be equal to 1. If there is absolutely no correlation between the wind speed at time t and the speed measured at the time $(t + \Delta t)$ then $R = 0$. For actual wind measurements the correlation function expressed as a function of lag time, Δt, will probably look something like the curve shown in figure 10.3. For $\Delta t = 0$ we obtain $R = 1$. From there on, the curve falls off more or less rapidly towards zero, depending on the size and prominence of eddies that make up the spectrum of atmospheric motions.

Equation 10.5 or figure 10.3 does not yet give us the 'spectrum' of atmospheric motions in terms of eddy size measured by the wave number

$$k = \frac{2\pi}{\Delta x} = \frac{2\pi}{\bar{V}\Delta t} \tag{10.6}$$

or frequency (cycles per second) at which certain eddies pass over a station. Equation 10.5 or figure 10.3 rather estimate the *relative* importance of eddies whose size is measured by a certain lag time, Δt, in expressing the ragged wind trace shown in figure 10.2. A mathematical trick, called Fourier transformation,

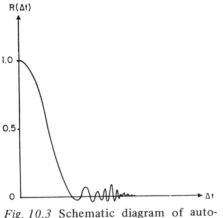

Fig. 10.3 Schematic diagram of auto-correlation function.

lets us obtain the 'spectrum', S(f), i.e. the eddy kinetic energy as a function of frequency, f, from the autocorrelation function $R(\Delta t)$. In essence, the spectrum function provides a 'Fourier analysis' of the wind field, giving us the *mean amplitudes* (i.e. a measure of kinetic energy) of sine- or cosine-shaped eddies of all possible frequencies contained in the time space $T/10$ or $T/5$.

The summation of all these kinetic energies S(f), also called 'spectral density', over all possible frequencies covered by our measurements has to yield the total variance, $\overline{V'^2}$, of the data sample shown in figure 10.2.

The meaning of equation 10.6 may also be expressed in the following way: S(f) expresses the kinetic energy of the spectrum band between k and $k + \Delta k$, or according to equations 10.6 and 10.4, between the lag times Δt and $\Delta t + \delta t$, where δt expresses a commensurate small increment in lag time. Thus, S(f) becomes the kinetic energy equivalent of the intensity of light in a certain region of the spectrum.

Let us finally look at a typical distribution of the spectrum function in the atmosphere. Vinnichenko (1970) compiled the data from which figure 10.4 was constructed. It shows the kinetic energy of eddies, expressed in the form of the spectrum function S(f), for all eddies measuring a few centimetres in diameter to eddies whose equivalent size expresses the variation of atmospheric motions in a five-year period. The spectrum, for the sake of convenience, has been plotted in figure 10.4 on logarithmic scales, because both the wave numbers considered, as well as the associated kinetic energies in the respective wave-number bands, vary over several orders of magnitude.

The units in which S(f) is expressed in this diagram are (km^2 h^{-2}) x day, which are the dimensions of kinetic energy per unit mass and per unit bandwidth of frequency, the latter being given in terms of cycles per day. Note that the scale for the spectra of the N–S wind component has been shifted relative to that of the E–W wind component, in order to avoid a confusing overlap of lines and shades areas.

The results shown in figure 10.4 pertain to measurements in the free atmosphere near an altitude of 10 km. In principle, measurements near the ground reveal similar spectra.

From this figure, we see that the energy of atmospheric motions, measured by the spectral density, increases with increasing eddy size (or decreasing frequency), at least up to wind fluctuations commensurate with a one-month periodicity. We also see from figure 10.4 that, depending on the meteorological circumstances, the kinetic energy in any one wave band within the spectrum may vary over more than one order of magnitude. For small eddies of a duration of less than one minute this variability may actually be as large as three orders of magnitude. This expresses the fact that the state of turbulence in the free atmosphere varies between almost laminar flow conditions under which birds, sailplanes and jet aircraft experience an extremely smooth ride, and very rough conditions, such as in thunderstorms, which may even result in structural damage to aircraft.

The foregoing example may suffice to show that spectrum analysis of atmospheric motions offers a powerful tool in describing statistically the behaviour of

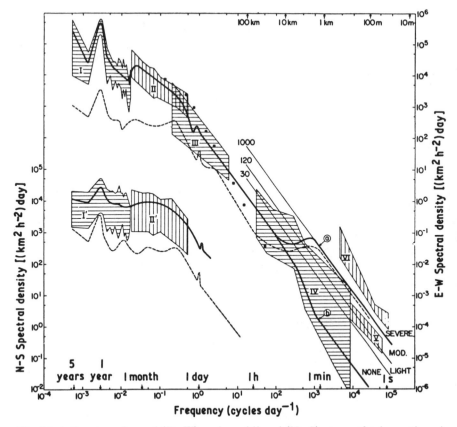

Fig. 10.4 Spectra of zonal (E—W) and meridional (N—S) atmospheric motions in the upper troposphere. Shaded areas, letters and roman numerals, as well as dots and lines, indicate different sets of measurements and their order-of-magnitude spread. The high-frequency end of the spectra (near periods of one to several seconds) is characteristic of clear-air turbulence. Note the increase in spectral density between no, light, moderate and severe turbulence. Area VI is characteristic of thunderstorm turbulence. The spectral peak at a period of 1 year indicates seasonal changes of atmospheric flow. Note that the scales for zonal and meridional flow are offset; nevertheless, there is more variability on the seasonal scale in the zonal flow than in the meridional flow (after Vinnichenko, 1970).

the atmosphere over a wide range of scales. The engineering applications of such analyses are obvious. If we know, for instance, the resonance frequencies to which a certain building, structure, aircraft frame, etc., is susceptible, spectrum analyses of atmospheric turbulent motions can tell us immediately how much kinetic energy the atmosphere is capable of providing in these critical frequency ranges, and whether or not the structure under investigation is in danger of damage or destruction.

References

Vinnichenko, N. K. (1970) 'The kinetic energy spectrum in the free atmosphere — one second to five years', *Tellus*, 22, 158—66.

Suggested reading

Sutton, O. G. (1953) *Micrometeorology*, New York, McGraw-Hill, 88—103.

11
Some aspects of the description of atmospheric turbulence

A. IBBETSON
University of Reading

Turbulence in the atmosphere exerts a very important influence on our environment. For example, at a time when we are becoming increasingly aware of the effects of pollution, both in the atmosphere and in the ocean, it is natural to try to elucidate the mechanisms responsible for its dispersal. Of course, this is only one of many fields in which turbulence plays a dominant role: the microclimate of all growing plants is controlled largely by the turbulent flow of air around them. Turbulence provides rapid and efficient vertical mixing of carbon dioxide and water vapour which would not be achieved in smooth flow, and the thermal environment of the plants is controlled by the redistribution of heat energy from the Earth's surface by the turbulence. In fact, the efficient mixing of tracers (whether particles, gases or physical properties of the air) in a turbulent flow is perhaps one of its most obvious characteristics. Calculation of the dispersion rate of tracer particles released into a turbulent flow is one of the simpler problems which can be approached theoretically. Prediction of the effects of the wind on man-made structures also demands an understanding of the properties of turbulence: the way in which a structure responds to the wind depends on the way in which the turbulent energy of the wind is distributed amongst eddies of different sizes.

Turbulence is unfortunately one of the few fields in science where theory lags far behind experiment. The rapid increase in our experimental knowledge of atmospheric turbulence in the last few decades is largely due to the development of sophisticated instruments which can measure its properties, and of electronic computers which are capable of dealing with large amounts of data. Even with these recent technical advances, however, the problem of making reliable measurements is far from trivial.

The purpose of this chapter is to introduce the reader to some of the basic ideas and terminology used in turbulence and to provide simple examples of the types of turbulent flow arising in the atmospheric boundary layer near the ground. The illustrations given here are mainly concerned with the 'ideal' case of air flowing over a flat, uniform plane – a gross simplification of the real atmospheric boundary layer. Even this relatively simple case, although it has been much studied recently, is not yet fully documented or understood. A further example of a more disturbed type of flow (which is effected by large obstacles and topography) is included as a reminder that, in very few applications over land, can conditions be regarded as ideal (although the boundary layer over the sea surface more often approaches ideal conditions).

Origins of turbulence

Some readers may be familiar with the experiments of Osborne Reynolds, in which water flowed through a straight tube of circular section and the motion of the water was revealed by injecting a continuous stream of dye along the tube. These and later experiments demonstrated that the nature of the flow through the tube changes dramatically as the flow speed is increased beyond a certain critical value (which depends on the tube diameter and the viscosity of the water). Above the critical speed the flow (as revealed by the injected dye) ceases

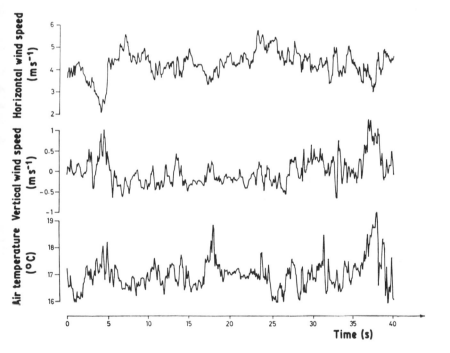

Fig. 11.1 A short section of a record of horizontal wind speed, vertical wind speed and temperature at a height of two metres at the meteorological field site at Reading University. Note the positive correlations between vertical wind speed and temperature, and the negative correlations between horizontal and vertical wind speeds.

to be in straight, parallel lines (laminar) as it was at lower speeds, becoming sinuous and eddying (turbulent). The turbulent flow is characterized by very rapid mixing of the dye filament across the whole diameter of the tube, whereas in laminar flow the mixing of the dye filament into the surrounding clear water is negligible.

In a similar way, the flow of air in the atmospheric boundary layer (the lower one or two kilometres of the atmosphere) tends to be turbulent when the wind speed is high enough and lapse conditions exist. The turbulent flow in the boundary layer always arises as the result of an initial instability of a laminar flow, but we are not concerned with that process here. Once initiated, the turbulence can be maintained against the dissipative, damping action of viscosity by two mechanisms: (1) shear of the wind, i.e. spatial variations of the average wind velocity; (2) thermal convection, i.e. the partially organized motion of buoyant air parcels upwards and the compensating motion of cooler air downwards. These two mechanisms give rise to turbulence having somewhat different properties, but we shall examine some features common to both.

The shear of the wind in the boundary layer is usually produced by frictional retardation, resulting either from flow over the ground (where the wind speed is necessarily small) leading to vertical wind shear, or from flow past obstacles (trees, buildings, or topographic features), in which case the wind variation may be in the horizontal. Either circumstance can lead to turbulent flow and, if the shear is maintained externally, the turbulence is able to increase its kinetic energy at the expense of the larger-scale average air motion. The sizes of the turbulent eddies in a turbulent shear flow are usually limited by the size of the region over which the shear exists: e.g., a building produces eddies not much larger than its largest dimension.

The convectively produced turbulence in the boundary layer is caused by heating of the ground by solar radiation. Warmer parcels of air rising from the ground eventually organize themselves into thermals (whose structure will depend on any wind present). The largest eddies produced in convective motions are roughly the same size as the depth of the boundary layer (typically 1 km), although near the ground characteristic eddy sizes are much smaller.

General properties of turbulence

Figure 11.1 shows a short section of a record of a turbulent flow measured at a height of 2 m in the meteorological field site at Reading University. (This is by no means an 'ideal' site, surrounded as it is by high trees and buildings at distances no greater than 200 m.) The records of vertical and horizontal wind speed were obtained using heated-wire anemometers (which respond very rapidly to gusts), while the temperature record was obtained using a fine-wire resistance thermometer. Each sensor produces a voltage which is proportional to the quantity it measures, and which can either be processed electronically during the experiment or recorded for later use. We can see the following characteristics in the records.

Each variable shows a degree of randomness, but there is also evidence of structure in the turbulent flow: over intervals of several tens of seconds we see large quasi-periodic changes in wind speed components and in temperature — evidence of large eddies in the flow.

Superimposed on the large slow changes are smaller, more rapid fluctuations: these, on closer examination, would show still smaller-scale structure on even shorter time scales. (The upper limit to the frequency of oscillations in the record is governed by the speed of response of the recorder.) It is apparent that the amplitudes of the longer period fluctuations are larger than those of shorter periods — a fact which will be verified later by quantitative analysis. Again, it is fairly obvious that, although there may be some relationship between values of, say, temperature measured a few seconds apart, the relationship will become less pronounced over longer intervals of time — a fact which we can confirm more precisely later by measuring correlations.

There is also some relationship between the records of the velocity components and the temperature: on average, increasing deviations of temperature T and upward vertical velocity w tend to occur together, as do the negative temperature deviations and downward vertical velocity. Thus, there is an apparent correlation between the variables w and T. Since positive deviations of w and T imply the upward motion of warmer air, while negative values imply downward motion of cooler air, the magnitude of the correlation is a measure of the rate of heat transfer upwards from the surface.

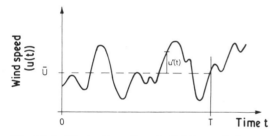

Fig. 11.2 Illustrating the definition of the mean wind speed, $\bar{U}$, over a period T and the fluctuations, $u'(t)$, from that mean.

There is also a tendency for periods of upward vertical motion to be associated with a decrease in the horizontal wind speed, while downward vertical motion tends to coincide with an increase in the horizontal wind speed. On average, then, faster moving parcels of air tend to be moving downward, implying that there is a net transfer of the horizontal momentum of the air downward by the turbulent eddies. This is the mechanism which enables the air to exert a drag force on the ground.

Given records such as those shown in figure 11.1, it is obvious that any description of the structure of the turbulence must be a statistical one: it would be unreasonable to think of a 'deterministic' description of variables showing so much randomness.

Statistical description of turbulence

We must first decide which types of variations in the flow will be classified as 'turbulence': we obviously wish to include all fluctuations which are rapid enough, but to exclude the slower variations characteristic of the larger scales of motion (mesoscale and synoptic scales) having periods longer than, say, ten minutes. Over an interval of time, T, therefore, which is usually a few minutes, we can measure the average wind direction and define an average wind speed $\bar{U}$ along that direction as

$$\bar{U} = \frac{1}{T} \int_{t-T/2}^{t+T/2} u(t)dt,$$

where $u(t)$ is the wind speed component along the average wind direction at time t. The instantaneous wind speed component $u(t)$ deviates from the average speed $\bar{U}$ by an amount

$$u'(t) = u(t) - \bar{U};$$

the turbulent fluctuations u' are evidently distributed equally above and below the mean, $\bar{U}$ (see figure 11.2).

The variability of the wind speed is best defined in terms of the average value of the squares of the deviations from the mean, i.e. the variance $\sigma_u^2 = \overline{u'^2}$, where the bar over the symbol denotes the average value over a period T. Turbulence is three-dimensional, so there will also be fluctuations in the wind components in the two directions at right angles to the mean wind: the horizontal component v' (at right angles to the mean wind) and the vertical component w', each of which has its own variance, σ_v^2 and σ_w^2. A useful measure of the 'gustiness' of the wind is the turbulence intensity, defined for each direction as $I = \sigma/\bar{U}$, σ being the square root of the variance (the standard deviation) of the appropriate wind component.

Within the boundary layer at a fixed height, the variances generally increase with increasing wind speed, and also with increasing convective activity. However, since the variances vary less rapidly with height above the ground than the average wind speed, the turbulence intensity, I, usually decreases with increasing height.

Values of I near the ground for the horizontal component u' are typically about 0.2–0.25 in daytime, while at night, I may decrease (as the air near the ground becomes more stably stratified). The corresponding value of I for the results shown in figure 11.1 is about 0.3, the difference presumably being due to the disturbed flow over the Reading site.

In order to say more about the structure of the turbulence, we need to be able to specify how much eddies of different sizes contribute to the total turbulent flow. However, it is very difficult to make measurements which give direct information about eddy sizes: measurements of turbulent quantities are usually made at a fixed point in space over a period of time. There is no simple general relation between measurements made in this way (as a 'time series') and the description of the spatial distribution of the eddies, because the eddies interact and develop with time in the turbulent flow. However, if the turbulent flow patterns can be regarded as slowly changing, a simple relation does exist between the temporal and the spatial descriptions. In this case, the turbulence is assumed to be 'frozen' into a fixed pattern which then moves along with the average wind speed. Changes with time seen at a fixed point in space are then identical to (although reversed in direction from) the spatial variations seen at a given instant of time: the wind advects each eddy past the point of observation. This is illustrated schematically in figure 11.3. A 'frequency' n observed in the temporal description is thus related to a 'wavelength' λ in the spatial description through the usual wave equation $n = \bar{U}/\lambda$. Although Taylor's 'frozen turbulence' model is found to be a good approximation for the smaller eddies of the turbulent flow, it appears to be an inadequate representation of the behaviour of the larger eddies (wavelengths greater than about 100 m), whose speed of translation is not always equal to the local average wind speed.

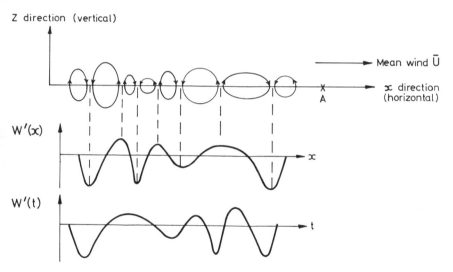

Fig. 11.3 A vertical cross-section through an idealized pattern of eddies advected past an observer at A. His description of the distribution of vertical velocity in space $w'(x)$ is just the exact reverse of his description of vertical velocity in time $w'(t)$.

Autocorrelations

One possible way of describing the statistical time variability of the flow is through the *autocorrelation* of the wind speed. For the wind component u' along the mean wind direction, the autocorrelation $R(\tau)$ is calculated from the product of the velocity fluctuation at any instant $u'(t)$ and its value a time τ later: $u'(t + \tau)$. This product is calculated at each successive instant of time, the average value of the product is calculated over a suitable length of time (usually several minutes), and is then divided by the total variance σ^2. Thus, we may write

$$R(\tau) = \overline{u'(t)u'(t + \tau)} \, / \, \overline{u'^2},$$

provided that the variance of u' does not change appreciably with time. The maximum possible value of R is evidently 1 (when $\tau = 0$), and we expect R to decrease as the time lag, τ, increases.

The shape of the autocorrelation curve $R(\tau)$ obviously depends on the structure of the turbulence: consider a flow in which only small eddies, randomly distributed in space, are moved past an observer with a given wind speed (cf. figure 11.3): the autocorrelation R will decrease more rapidly with τ than if the flow contained only large eddies. Provided $R(\tau)$ eventually decreases to zero at large

values of τ, the area under the whole autocorrelation curve $\int_0^\infty R(\tau)d\tau$ is therefore

a measure of the period associated with the largest eddies in the turbulence, and is usually called the integral or outer scale of the turbulence, denoted T_I. Using Taylor's 'frozen turbulence' model, the physical size of the largest eddies would therefore be $\bar{U}T_I$.

In practice, the autocorrelation is usually estimated numerically by first *digitizing* the continuous record of wind speed: the value of the wind speed is read from the record at suitable intervals of time, and R is then calculated from these discrete values, as illustrated in figure 11.4.

The autocorrelations of wind components v' and w' will, in general, be different from that of u', although if the turbulence has statistically identical properties in all three directions (isotropic) there will be simple relations between them.

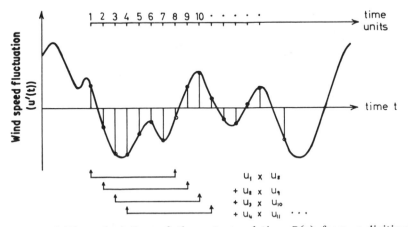

Fig. 11.4 The calculation of the autocorrelation $R(\tau)$ from a digitized wind speed fluctuation record (circles on the continuous curve): e.g. the value of $R(\tau)$ at a time lag, τ, of seven time steps is obtained from $u'_1 \times u'_8 + u'_2 \times u'_9 + u'_3 \times u'_{10} + \cdots$ divided by the total variance $\sigma^2 = u'^2_1 + u'^2_2 + u'^2_3 + \cdots$.

Typical autocorrelations for turbulence in the boundary layer are shown in figure 11.5: curves A and B (u' and w') refer to a height of 2 m above open level ground under lapse conditions, while curve C refers to the horizontal wind speed record shown in figure 11.1. It is worth noting that the autocorrelations for the horizontal and vertical components, curves A and B, differ by much more than they would if the turbulence were isotropic.

Fourier analysis and energy spectra

The autocorrelation thus contains information about the structure of the turbulence: however, it is not easy to deduce from it immediately how much different eddy sizes contribute to the total variance of a turbulent wind component in a given direction. In order to do this, we must be able to select a given eddy size (or frequency – they are related in Taylor's frozen turbulence model) and measure the variance associated with it. In optics, a beam of light can be split up into a spectrum of colours of different wavelength, each band of wavelengths having its own intensity (or energy). By analogy, the result of splitting up the total variance of a turbulent velocity component into contributions from different eddy sizes is also called a spectrum. Since the variance of the velocity is proportional to the turbulent kinetic energy per unit mass of air, the spectrum of the variance is simply proportional to the spectrum of kinetic energy.

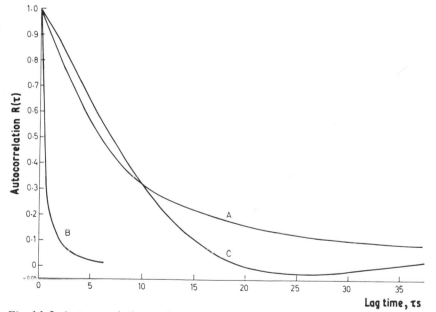

Fig. 11.5 Autocorrelations observed in the boundary layer at a height of two metres. Curves A and B show autocorrelations of horizontal wind speed u' and vertical wind speed w' over open flat country. Curve C is the autocorrelation of the horizontal wind speed shown in figure 11.1.

The technique by which the variance associated with a given eddy size can be calculated is Fourier analysis. This is based on the principle that any quantity which varies periodically with time may be expressed as the sum of a number of Fourier components (sine and cosine waves) of different frequencies, each having its own amplitude. For example, a square wave $f(t)$ of fundamental frequency n_0 and unit amplitude (figure 11.6) may be expressed as the sum of an infinite number of sine wave components of frequencies n_0, $3n_0$, $5n_0$ . . . etc., of decreasing amplitudes:

$$f(t) = \frac{4}{\pi} \left(\sin 2\pi n_0 t + \frac{1}{3} \sin 6\pi n_0 t + \frac{1}{5} \sin 10\pi n_0 t + \cdots \right).$$

Figure 11.6 also shows the 'spectrum' of the square wave: it is a *line* spectrum because the Fourier components are all multiples of the fundamental frequency n_0: the height of each line is just the square of the amplitude of the Fourier component. (Of course, the spectrum of a pure sine wave of frequency n_0 is a single line at that frequency.)

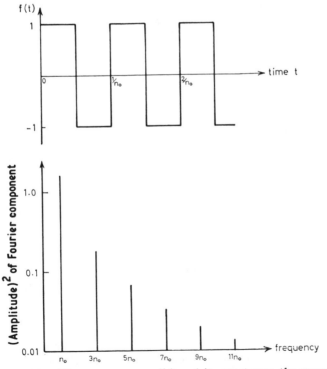

Fig. 11.6 The square wave $f(t)$ and its spectrum: the square of the amplitude of each Fourier component is plotted logarithmically as a function of frequency.

Similarly, using Fourier analysis techniques, we can analyse a non-periodic time-dependent variable, such as a turbulent wind speed component. In this case, if the duration of the record is indefinitely long, the number of Fourier components required to represent the variable is infinite, and all possible frequencies from zero to infinity must be included. The spectrum is no longer a line spectrum, but becomes a continuous distribution function – the energy spectrum $E(n)$.* This is defined so that $E(n)dn$ is the variance associated with Fourier components whose frequencies lie in a frequency interval dn (from n to $n + dn$). Summed over the whole possible range of frequencies, the total variance must be σ^2:

$$\sigma^2 = \int_0^\infty E(n)dn.$$

The energy spectrum of, say, a wind speed component may in practice be estimated either electronically or numerically. The electronic method uses electrical filters (each of which transmits electrical signals only in a narrow range of frequencies, rejecting all other frequencies), to separate out the variations in the wind record in the frequency range of interest: the variance in that frequency range is then measured electronically by squaring and averaging the output from the filter. The numerical methods again use digitized information, i.e. the continuous record is sampled at suitable time intervals. In the simplest method, the spectrum is then obtained numerically from Fourier analysis of the sampled data points. Since the data are discrete points and the length of the record is necessarily finite, the number of Fourier coefficients is also finite. The resulting line spectrum is then averaged over suitable frequency bands to provide an approximation to the actual (continuous) energy spectrum.

Energy spectra may therefore be defined for each of the three wind components u', v', and w' as above. As these spectra are obtained from a one-dimensional Fourier analysis of a wind component, they should strictly be called one-dimensional energy spectra. Since the Fourier analysis is carried out with respect to frequencies measured along the mean wind direction, the wind components v' and w', being transverse to that direction, behave differently from the component u', which is along the mean wind direction. The one-dimensional energy spectra of v' and w' are therefore different from that of u', even in isotropic turbulence.

Typical 'ideal' boundary layer spectra of the vertical wind component, w', and the horizontal wind component (along the mean wind direction), u', are shown in figure 11.7. Before discussing them, it is appropriate to comment on the formats commonly used in these graphs. Conventionally, the energy spectrum $E(n)$ is plotted as a function of frequency n, both axes having logarithmic scales as in figure 11.7a. This has the advantage of allowing presentation of results over a wide range of frequencies, but it distorts the shape of the spectrum. (It must be remembered that $E(n)dn$ is the contribution to the variance at frequency n,

---

* There is no agreed notation: $S(n)$ and $\theta(n)$ are used also to denote the energy spectrum.

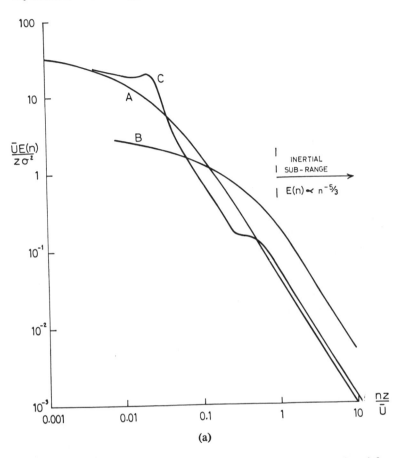

Fig. 11.7(a) The energy spectrum $E(n)$ plotted in conventional format against frequency n, both axes having logarithmic scales.

so on a logarithmic frequency plot a constant frequency interval dn has an increasing linear width as frequency decreases.)

Some meteorologists prefer to plot $nE(n)$ linearly versus frequency n on a logarithmic scale, as in figure 11.7b. Since the contribution to the variance in a bandwidth dn is then

$$E(n)dn = nE(n)\frac{dn}{n} = nE(n)\,d(\ln n),$$

this format implies that equal areas under the graph represent equal contributions to the variance. However, the shape of the spectrum is again distorted in this type of graph, so any interpretation of the spectral shape in either the

148

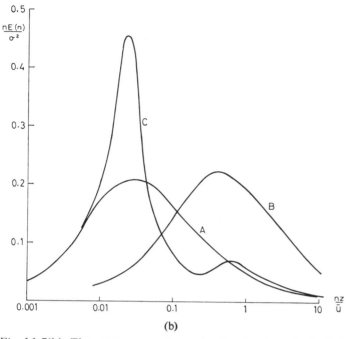

Fig. 11.7(b) The energy spectrum in the 'meteorological' format: $nE(n)$ is plotted linearly against frequency on a logarithmic scale. The normalization of the ordinates and the abscissa are explained in the text. Curves A and B are energy spectra of horizontal wind speed u' and vertical wind speed w' in an ideal boundary layer over open flat country. Curve C is the spectrum of horizontal wind speed over the meteorological site at Reading. Note the presence of anomalously energetic large eddies (wavelengths about 80 m) in this very disturbed flow.

'conventional' or the 'meteorological' format must return to the basic definition of $E(n)$.

As discussed earlier the frequency n observed at a particular point may be regarded as due to the motion of an eddy of wavelength λ past that point with the mean wind speed $\bar{U}$. Thus, spectra measured in different wind speeds have similar shapes, but the frequencies observed increase with the wind speed: $n \propto \bar{U}$. Similarly, spectra measured at different heights above ground have similar shapes, but the frequencies observed depend on height: the wavelengths of the eddies tend to be larger at greater heights. Thus λ increases with height z, so n increases with $1/z$. This suggests that spectra plotted with $\log nz/\bar{U}$ as abscissa may show less scatter between data obtained at different heights and wind speeds. $nz/\bar{U}$ is commonly referred to as the 'reduced frequency'.

It is also usual to 'normalize' the ordinates of these graphs to make them dimensionless (pure numbers): in the meteorological format, $nE(n)$ is divided by

σ^2, the total variance. The total area under the graph of $nE(n)/\sigma^2$ versus log $nz/\bar{U}$ (figure 11.7b) is then unity by definition. In the conventional format, $E(n)$ is divided by σ^2 and a time scale to make it dimensionless. The time scale chosen is $z/\bar{U}$ (for the reasons given in the last paragraph). We therefore plot log $\bar{U}E(n)/z\sigma^2$ versus log $nz/\bar{U}$ in figure 11.7a.

Returning to the energy spectra shown in figure 11.7 we see that in the 'ideal' boundary layer, u' and w' have spectra of similar shape (curves A and B). In the meteorological format (figure 11.7b) they show maxima at reduced frequencies of about 0.3 (for w') and 0.03 (for u'), although these values are found to depend on the local lapse rate. The low frequency part of the u' spectrum depends very much upon the details of the large-scale flow, e.g. compare A, which refers to flow over open flat country, with C, which is the spectrum of the horizontal wind speed shown in figure 11.1 measured in Reading. Here, the large eddies (low frequencies) contribute much more to the variance than in the flow over open country. The maximum variance in curve C is associated with wavelengths of about 80 m, presumably an indication of the characteristic scale of the obstacles disturbing the flow over the Reading field site.

On the other hand, the spectra at high frequencies ($nz/\bar{U}$ greater than about 1 in figure 11.7a) show a great similarity in all cases, an observation which merits further comment. This arises because turbulent kinetic energy is generated (from shear or from buoyant convection) principally at larger scales (lower frequencies). These large eddies interact, producing smaller eddies which then interact, and so on. Energy is therefore transferred from larger to smaller eddies in a continuous cascade. Ultimately, the energy is transferred to such small eddies that the effects of molecular friction (viscosity) can no longer be neglected, and the kinetic energy of the turbulence is dissipated in these small (high frequency) eddies, reappearing as a small heating of the air. This part of the spectrum, where viscosity is important, is known as the 'viscous sub-range' of frequencies. In the intermediate range of eddy sizes (or frequencies) where the eddies have lost any 'memory' of how they originated, but have not become so small that friction is important, the transfer of energy from larger to smaller scales of motion is the dominant physical process. This part of the spectrum is known as the 'inertial sub-range' of frequencies, and it is in this sub-range that Kolmogorov predicted the form of the energy spectrum to be

$$E(n) \propto n^{-5/3},$$

the well known 'minus five-thirds law'. Both of these sub-ranges of the spectrum are referred to as 'universal', implying that the statistical properties of the turbulence at these small wavelengths are not dependent on how the turbulence originated (whether from shear or from convection), because sufficient eddy interactions have occurred since its generation that the flow has lost any memory of how the original eddies (of larger wavelengths) were produced.

It might be expected using the simplest physical reasoning that there should be a relation between energy spectra and autocorrelations: the value of the autocorrelation at large time lag, τ, is determined mainly by the structure of the larger eddies of the turbulence, as is the value of the energy spectrum at low frequencies, n. This relationship can be expressed mathematically: the energy spectrum, $E(n)$, is the Fourier cosine transform of the corresponding *autocovariance* $\sigma^2 R(\tau)$, and vice versa. Further discussion of this relationship unfortunately involves more mathematical complexity, and so is beyond the scope of this chapter, but its significance can be appreciated: experimenters need to measure only *one* of these functions (at least in principle) in order to describe the turbulence.

Discussion

We are now in a position to look in more detail at one of the examples mentioned in the introduction: the dispersion of pollution in a turbulent atmosphere. In considering the general problem of how material is dispersed by turbulence, it is evident that we need first to be able to calculate the cumulative effect of all the turbulent eddies in a flow on a small volume of marked particles released at a point in a turbulent wind. In particular, we need to know how much the particles will spread in the two cross-wind directions (vertically and laterally) as they travel downwind. G. I. Taylor first gave the solution to this problem in terms of the autocorrelation function of the turbulent motion, but it is better to re-interpret it in terms of the energy spectrum. Thus, it is found that the width of the cross-wind spread at some place downwind of the source is equal to the product of the time taken for the particles to travel there and the square root of the variance (the standard deviation) of the turbulent velocity component in that cross-wind direction. However, the variance is obtained by summing over only a limited range of the energy spectrum, from the lowest possible frequencies to an upper 'cut-off' frequency which decreases as the time of travel of the particles increases, i.e. as we go further from the source. Thus, the dispersion at large distances from the source is determined mainly by the larger eddies in the flow and the smaller, high-frequency eddies do not contribute to the permanent lateral displacement of a particle although they are, of course, still present in the turbulence.

Although we have considered only a point source of pollution here, the extension of these ideas to larger, more complex sources presents no conceptual difficulties. In practice, of course, it is not possible to use this simple method as a means of predicting dispersion, as too many important meteorological factors have been omitted (the shear of the wind in the boundary layer, and the effect of static stability on vertical dispersion, among others). However, this simple model does show how the real problem must be tackled, and its essentials are incorporated in one of the practical methods of predicting atmospheric dispersion (see e.g. Pasquill, 1962).

This chapter was intended to provide only an introduction to some of the simpler aspects of the description of turbulent flow, without looking at its dynamics. Hopefully, it has also indicated how information about the structure of turbulence can assist in solving some of the environmental problems with which meteorologists must be concerned. Unfortunately, any further examination of turbulence must involve more mathematics, but interested readers may like to sample the literature noted below, none of which is too demanding mathematically.

Acknowledgements

I should like to thank Mrs K. Daykin for her preparation of the diagrams, Dr M. A. Pedder and Mr G. K. Thwaites for assistance in computing spectra, and Dr J. R. Milford for his helpful comments on the manuscript.

Reference

Pasquill, F. (1962) *Atmospheric Diffusion*, New York, Van Nostrand.

Other reading

Owen, P. R. (1971) 'Buildings in the wind' (Symons Memorial Lecture), *Quart. J. R. Met. Soc.*, 97, 396–413.
Sutton, O. G. (1949) *Atmospheric Turbulence*, London, Methuen.

12
Atmospheric energetics

A. A. WHITE
Meteorological Office, Bracknell

The earth's atmosphere is a very complicated physical system, yet the laws which govern its behaviour are amongst the longest established and best loved in physics. It is not surprising, therefore, that some of the fundamental concepts of classical physics are important in atmospheric dynamics. In this chapter one such concept, *energy*, is considered. The energetics of a system consisting of a single particle are investigated first, in order to recall basic ideas. Then we deal with the extension to fluid systems, and finally various atmospheric energy cycles are discussed. Our object throughout is only to outline these various aspects of energy theory. As far as possible, mathematical obscurity will be avoided.

Energy in a single particle system

One gramme of air contains about 2×10^{22} molecules, and the entire atmosphere about 10^{44}. Before tackling the energetics of the atmosphere, we shall introduce the essentials of energy theory by examining a somewhat simpler system – a single particle. The main concepts are kinetic energy, the rate of working of a force, and potential energy.

Kinetic energy *(KE)*

Figure 12.1 depicts the motion in two dimensions, y, z, of a particle of constant mass, m. At some instant, its velocity components in the directions y, z are v, w respectively and it is subject to a force $\mathbf{F}$ whose components are F_y, F_z. Application of Newton's Second Law of Motion (see Atkinson, 1981) in the y and z directions gives

$$m\frac{dv}{dt} = F_y, \quad (12.1) \qquad m\frac{dw}{dt} = F_z. \qquad (12.2)$$

Multiplying equation 12.1 by v, equation 12.2 by w and adding the results we find that

$$\frac{d}{dt}\left\{ \frac{1}{2}m(v^2 + w^2) \right\} = vF_y + wF_z. \tag{12.3}$$

The quantity $\frac{1}{2}m(v^2 + w^2)$ is called the kinetic energy (*KE*) of the particle. The quantity $vF_y + wF_z$ (in which each velocity component is multiplied by the corresponding force component) is called the *rate of working of the force* **F**. Equation 12.3 states that the rate of change of the particle's *KE* is equal to the rate of working of the force **F** acting on it. A result analogous to equation 12.3, and similar definitions, apply to motion in three dimensions.

What is the use of equation 12.3? Its advantages (and limitations) will become apparent as we proceed. Here we note only two points: (1) *KE* is a function of the *speed* $(v^2 + w^2)^{1/2}$ of the particle, but is independent of the *direction* of its motion; (2) because it is a single equation, equation 12.3 contains less information than the two equations from which it was derived.

Points (1) and (2) reflect the fact that the state of motion of a particle is not completely specified by its *KE*.

Two cases in which equation 12.3 is applicable are of particular interest: motion subject to a force acting perpendicular to the instantaneous direction of motion, and motion under gravity.

Forces acting perpendicular to the motion

Suppose that the force **F** acts at right angles to the direction of motion of the particle. Referring to figure 12.1, it follows that

$$vF_y + wF_z = 0.$$

Thus, equation 12.3 becomes

$$\frac{d}{dt}\left\{ \frac{1}{2}m(v^2 + w^2) \right\} = 0 \tag{12.4}$$

in this case. From equation 12.4 we conclude that *the KE of a particle is not affected by forces which act at right angles to its direction of motion*. The result holds also for motion in three dimensions.

Here we have a further indication that the state of motion of a particle is not completely specified by its *KE*. Forces which act at right angles to the motion are obviously important in the dynamics of a system — as anyone who has ever entrusted his life to the more frightening machinery at a fairground will surely agree. Yet these forces do not figure in the energetics.

In meteorological dynamics, the most important force which acts at right angles to the motion is the Coriolis force, one of the forces which apparently act when velocities are measured relative to the rotating Earth or any other rotating frame (see Panofsky, 1981a). Coriolis forces do not contribute to *KE* equations such as equation 12.3.

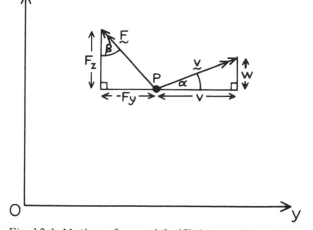

Fig. 12.1 Motion of a particle (*P*) in two dimensions *y*, *z*. The velocity **v** of the particle and the force **F** acting on it at some chosen time are shown vectorially. If **F** acts perpendicular to **v** then $\alpha = \beta$.

Hence $\tan \alpha = \tan \beta$, and, from the component triangles, this means that

$$\frac{w}{v} = -\frac{F_y}{F_z}$$

so that $vF_y + wF_z = 0$.

Note that the component triangle for F has its lower side of length $-F_y$ because F_y is negative for **F** acting in the direction shown.

Motion under gravity — potential energy (*PE*)

Consider the case in which the direction *z* is vertically upwards and gravity is the only force acting on the particle. Then $F_z = -mg$ and $F_y = 0$. Supposing that v is zero initially, it will remain so, and equation 12.3 can be written as

$$\frac{d}{dt} \left\{ \frac{1}{2} mw^2 + mgz \right\} = 0, \qquad (12.5)$$

since $w = dz/dt$, where *z* is the height of the particle above some reference level. Equation 12.5 is very significant. It states that *the quantity* $E = \frac{1}{2}mw^2 + mgz$ *remains constant during the particle's motion. mgz*, which clearly has the same dimensions as the *KE* (mass x length^2 x time^{-2}), is conveniently thought of as an energy term whose addition to the *KE* enables a constant total energy *E* to be defined. *mgz* is called the 'gravitational potential energy' or simply the '*potential energy*' (*PE*) of the particle. Upward motion is accompanied by conversion of *KE* to *PE*, downward motion by conversion of *PE* to *KE*. *PE* is really just an easy way of taking gravity into account.

What happens if forces other than gravity are also acting? It is convenient to re-define F_z to include the vertical components of all forces except gravity. The generalized form of equation 12.5, allowing force components also in the direction y, is

$$\frac{d}{dt}\left\{\frac{1}{2}m(v^2 + w^2) + mgz\right\} = vF_y + wF_z, \qquad (12.6)$$

which states that *the rate of change of the total energy* $E = \frac{1}{2}m(v^2 + w^2) + mgz$ *is equal to the rate of working of all forces other than gravity*. Extension to motion in three dimensions is straightforward.

Summary

The fundamental quantity in energy theory is kinetic energy (*KE*). Potential energy (*PE*) is simply a useful device for taking the force of gravity into account. The *KE* of a particle of constant mass is a measure (but not a linear measure) of how fast it is moving irrespective of its direction of motion. Thus a particle's state of motion is incompletely specified by its *KE*. The *KE* is changed only by forces which have a component parallel to the direction of motion. The special importance of *KE* is that, being a measure of speed, it is one of the most fundamental attributes of a particle in motion.

Most of the essential concepts of energy theory have now been discussed. How can they be applied to fluids?

Energy in fluid systems

In much the same way as for a single particle, energy relations for a fluid can be derived from the equations of fluid motion (see Panofsky, 1981a, b) by elementary algebraic operations. We shall discuss the results of this procedure later on. First, we shall introduce the major refinement of fluid energetics, namely the separation of the kinetic energy into the kinetic energy of bulk flow and the kinetic energy of the molecular motions – the *internal energy* (*IE*).

The total energy of a volume of fluid

The basis of fluid dynamics is the application of physical laws to small fluid elements which contain many millions of molecules. What is the total energy (*KE* + *PE*) of such an element? Clearly it is equal to the sum of the *KE*s and *PE*s of each molecule in the element. The total *PE* is that of the total mass of the molecules situated at their centre of mass. The total *KE* behaves differently, consisting of two parts: (1) the *KE* of bulk flow, which is the *KE* of the total mass of the molecules moving with the speed of their centre of mass; (2) the *KE* of the random thermal motions of the molecules, which is the sum of the *KE*s of the molecules measured relative to their centre of mass.

These general results are illustrated in figure 12.2 for the special case of two particles restricted to motion in a vertical direction.

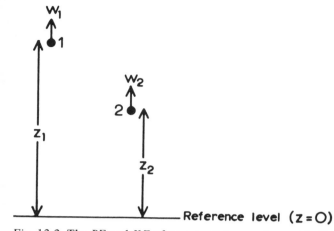

Reference level $(z = 0)$

Fig. 12.2 The PE and KE of two particles moving vertically. Particles 1 and 2, of equal mass m, have vertical velocities w_1 and w_2 and are at heights z_1 and z_2 at some instant. The total PE is

$$mgz_1 + mgz_2 = 2mg\left(\frac{z_1 + z_2}{2}\right),$$

and the total KE is

$$\tfrac{1}{2}mw_1^2 + \tfrac{1}{2}mw_2^2 = m\left(\frac{w_1 + w_2}{2}\right)^2 + m\left(\frac{w_1 - w_2}{2}\right)^2.$$

Thus the total PE is the same as that of the total mass $2m$ situated at the centre of mass $\tfrac{1}{2}(z_1 + z_2)$ of the system; but the total KE exceeds that of the total mass moving with the velocity $\tfrac{1}{2}(w_1 + w_2)$ of the centre of mass. The extra contribution to the total KE is equivalent to the total KE of particles 1 and 2 in terms of velocities measured in a co-ordinate system moving with the centre of mass. This simple argument is readily extended to the case of an arbitrary number of particles moving in three dimensions. In mathematical terms, the difference in behaviour of PE and KE arises because the PE of a particle is a linear function of its height whereas the KE is a quadratic function of its velocity components.

The KE of the random thermal motions is called the *internal energy* (*IE*). The contribution of the *IE* is clear in the special case of a fluid element at rest: the KE of bulk flow is then zero, but the element contains KE because of the random thermal motions of the molecules.

From now on we shall follow normal practice by using the term kinetic energy (*KE*) to indicate the kinetic energy of bulk flow. The total energy E of a

fluid element is

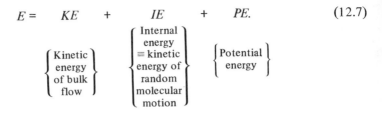

$$E = \quad KE \quad + \quad IE \quad + \quad PE. \tag{12.7}$$

If the mass of fluid in the element is M and its bulk flow (velocity) components in the (orthogonal) directions x, y, z are u, v, w respectively, the KE of the element is $\frac{1}{2}m(u^2 + v^2 + w^2)$. If z is the height of its centre of mass above some chosen reference level, its PE is Mgz. The IE of the element is simply Mc_vT, where c_v is the specific heat at constant volume and T is the absolute temperature. Thus equation 12.7 becomes

$$E = \frac{1}{2}M(u^2 + v^2 + w^2) + Mc_vT + Mgz. \tag{12.8}$$
$$\{KE\} \qquad\qquad \{IE\} \quad \{PE\}$$

Each of the three forms of energy corresponds to a property which is of natural interest. For a fixed mass M, the KE is a measure of the speed of flow, the IE is directly proportional to the absolute temperature and the PE is directly proportional to the elevation. In brief, the KE, IE and PE of a fluid element signify how fast, how hot, how high.

What are typical values of KE, IE and PE (per unit mass) in the atmosphere? Air speeds, V, relative to the rotating earth are a few m s^{-1} in cumulus circulation, about 10 m s^{-1} at low levels in an extra-tropical cyclone, and up to 100 m s^{-1} in a jet stream. The KE per unit mass $(= \frac{1}{2}V^2 = KE_1)$ is therefore generally in the range 0 to 5×10^3 J kg^{-1}. Since the specific heat at constant volume (c_v) for air is about 7×10^2 J kg^{-1} K^{-1}, the IE per unit mass $(= c_vT = IE_1)$ is generally in the range 14 to 21×10^4 J kg^{-1}. Taking $z = 0$ at ground level, the PE per unit mass $(= gz = PE_1)$ for air near the tropopause is about 10^5 J kg^{-1}, and 7×10^4 J kg^{-1} might be considered a reasonable average value for the atmosphere. PE_1 and IE_1 are thus typically of the same order, but even for extreme wind speeds (100 m s^{-1}), KE_1 is much less than IE_1 and typical values of PE_1. Later it will be seen that these important results can be at least partially accounted for.

Having identified the usual subdivisions of energy in a fluid system we shall next examine how the various kinds of energy can be generated, dissipated and interconverted. The balances of KE, IE and PE for an element of fluid as it flows are considered in the next section, and the balances for the entire volume of a fluid system are discussed in the remainder of the chapter. In both cases, the pressure field plays a complicating but crucial role, though there is a useful simplification in the second case.

The energy balance of a fluid element

For convenience we shall suppose the fluid element to have unit mass. To simplify the treatment further we shall consider motion in the vertical only. Generalization to the case of three-dimensional motion involves no new concepts – only somewhat longer equations.

Potential energy of unit mass (PE$_1$)

This is the simplest case. Since $PE_1 = gz$,

$$\frac{d}{dt}(PE_1) = gw. \tag{12.9}$$

PE_1 increases in upward flow and decreases in downward flow.

Kinetic energy of unit mass (KE$_1$)

Suppose that the volume of the unit mass element is α (the 'specific volume'), that p is the pressure and that Fr_z is the component of the frictional force in the direction z. From the vertical component of the momentum equation (see Panofsky, 1981a. Panofsky points out that Fr_z is negligible for almost all atmospheric motions; Fr_z is included here simply to make our one-dimensional problem typical of the general three-dimensional one) it is easily shown that KE_1 $(=\tfrac{1}{2}w^2)$ obeys the relation

$$\frac{d}{dt}(KE_1) = wFr_z - gw - \alpha w \frac{\partial p}{\partial z}. \tag{12.10}$$
$$\quad\text{(a)}\quad\quad\text{(b)}\quad\text{(c)}$$

Equation 12.10 states that the rate of change of KE_1 is equal to the sum of (a) the rate of working by friction; (b) the rate of working by gravity; and (c) the rate of working by the pressure-gradient force. Since its right-hand side involves only the rates of working of various forces, equation 12.10 is of the same general form as the equation for the rate of change of the KE of a particle (see equation 12.6).

Internal energy of unit mass (IE$_1$)

The internal energy balance must obey the First Law of Thermodynamics (see Panofsky 1981b). If the rate of heating per unit mass is Q, IE_1 $(=c_v T)$ must vary such that

$$\frac{d}{dt}(IE_1) = Q - p\frac{d\alpha}{dt}. \tag{12.11}$$

The term $-p d\alpha/dt$ represents the rate of working of the pressure field on the element *as a consequence of any rate of change of its volume.* When $d\alpha/dt < 0$

(compression of the element) the pressure in the surrounding fluid is working on the element so as to increase IE_1. When $d\alpha/dt > 0$ (expansion) the pressure in the element is working on the surrounding fluid so as to decrease IE_1. Equation 12.11 states that the rate of change of IE_1 is equal to the rate of heating per unit mass plus the rate of working of the pressure field in compressing the element.

We see that the pressure field can affect the energy balance of the element of fluid in two ways. Motion down/up a pressure *gradient* tends to increase/decrease KE_1. In compression/expansion the pressure field tends to increase/decrease IE_1. A little algebra shows that equation 12.11 can be rewritten in the form

$$\frac{d}{dt}(IE_1) = Q - \left\{ \alpha \frac{\partial}{\partial z}(wp) - \alpha w \frac{\partial p}{\partial z} \right\}. \tag{12.12}$$

(Equation 12.12 follows from equation 12.11 upon use of the appropriate continuity equation $d\alpha/dt = \alpha \partial w/\partial z$.) Reference to figure 12.3 shows that the term $-\alpha \partial/\partial z(wp)$ represents the total rate of working of the pressure field on the unit mass element. The term $-\alpha w \, \partial p/\partial z$ is the rate of working of the pressure-gradient force in producing KE_1. Thus, equation 12.12 shows that *the rate of working of the pressure field in producing IE_1 is equal to the total rate of*

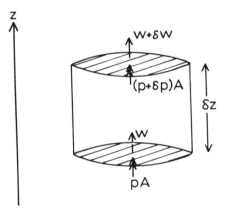

Fig. 12.3 The action of the pressure field on a unit mass fluid element (assuming motion in the z direction only). The element has height δz and circular cross-section A. At the lower and upper surfaces the pressure and vertical velocity are p and w, $p + \delta p$ and $w + \delta w$ respectively. At the lower surface the pressure in the underlying fluid is working on the element at a rate

vertical velocity x force in z direction $= wpA$.

At the upper surface of the element the pressure in the element is working on the overlying fluid at a rate $(w + \delta w)(p + \delta p) A$.

Thus, the total rate of working of the pressure field on the fluid element is $A[wp - (w + \delta w)(p + \delta p)] \simeq -A\delta z \, \partial/\partial z \, (wp) = -\alpha \, \partial/\partial z \, (wp)$, the latter equality holding because $A\delta z = \alpha$ (the volume of the unit mass element).

working of the pressure field on the element less the rate of working to produce KE_1.

Total energy

An equation for the total energy E_1 ($= KE_1 + IE_1 + PE_1$) of the element is obtained by adding equations 12.9, 12.10, 12.12:

$$\frac{dE_1}{dt} = Q + wFr_z - \alpha \frac{\partial}{\partial z} (wp). \tag{12.13}$$

E_1 changes only as a result of heating, total pressure work and work done by friction.

Mathematically, the simplicity of equation 12.13 arises because several of the terms on the right-hand sides of equations 12.9, 12.10 and 12.12 cancel on addition of these equations. The term $-gw$ in equation 12.10, representing the rate of working by gravity, appears with opposite sign in equation 12.9. The term $-\alpha w \, \partial p/\partial z$ in equation 12.10, representing the rate of working by the pressure-gradient force, appears with opposite sign in equation 12.12. Both $-gw$ and $-\alpha w \, \partial p/\partial z$ can therefore be regarded as *energy conversion terms*, since they represent a rate of change of KE_1 which is accompanied by an exactly equal and opposite rate of change of PE_1 and IE_1 respectively. We shall meet conversion terms in a slightly different context later.

Energy balances and conversions can be represented concisely by box diagrams. Figure 12.4 shows the diagram which represents equations 12.9, 12.10, 12.12 and 12.13 for the balances of a unit mass element. The relevant processes are heating and cooling, vertical motion, friction, motion up or down a pressure gradient, and the total pressure work.

What does the heating Q correspond to in the atmosphere? As noted by Panofsky (1981b), the major agencies of heating are radiative flux convergence, and latent heat release in condensation of water vapour; the reverse processes are agencies of cooling. Frictional dissipation (see figure 12.4), whereby bulk flow KE is 'degraded' into IE by molecular viscosity, contributes only to the heating. Molecular conduction can contribute heating or cooling, but is generally unimportant except very close to material surfaces. All these processes occur on the molecular scale and thus tally with the usual thermodynamic picture of heating. Frequently, however, Q is extended to include the convergences or divergences of fluxes carried by eddy motion occurring on a smaller space scale than that of the 'element' of air which is under consideration.

Our survey of the energy balances for a single fluid element is now complete. We now consider the energy balances for the entire volume of a fluid system — the balances formed by adding together the contributions of every fluid element in the system. We also discuss a succession of energy cycles. The most sophisticated cycle is very useful in characterizing the general circulation of a fluid system and also in testing numerical models. A convincing exposition of this important part of energy theory would involve rather complicated mathematics. Since we wish to avoid such complexities, our treatment must be superficial and

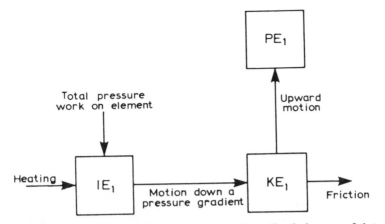

Fig. 12.4 Energy box diagram representing the balances of internal energy, potential energy and kinetic energy for a unit mass element of fluid. Each box represents a form of energy. Arrows joining two boxes denote energy conversion processes and are labelled according to the direction of the arrows: for the reverse processes the arrows would point in the opposite directions. Arrows connected to only one box represent other processes. The arrows representing heating and total pressure work are directed as indicated by the labels: they would also be oppositely directed for the reverse processes. Friction can act to increase or decrease KE_1, and so the relevant arrow may point in either direction, but in either case some frictional dissipation occurs which contributes to the heating. This is because any increase of KE_1 due to frictional interaction with surrounding fluid is accompanied by a larger decrease in the *KE* of the surrounding fluid. The deficit of *KE* appears as *IE*, or, in other words, *IE* is formed by dissipation of *KE*. Frictional dissipation is invariably regarded as part of the heating rather than as a conversion process. Amongst several good reasons for this is that it differs from the recognized conversion processes in being irreversible.

a number of crucial results will be stated without any semblance of proof. Nevertheless we shall try to convey the physical notions which underlie the detailed theory.

The global energy balance of a fluid system

To avoid confusion with other uses of the word 'total', the *KE, IE* and *PE* contained in an entire fluid system will be called the *global KE, IE* and *PE*. Corresponding to these global amounts, it is clearly possible to define global values of the various generation and conversion rates.

The global *KE, IE* and *PE* are conveniently indicated as $\overline{KE}$, $\overline{IE}$ and $\overline{PE}$. Equations for the rate of change of $\overline{PE}$, $\overline{IE}$ and $\overline{KE}$ can be derived from parcel balance equations (such as the three-dimensional analogues of equations 12.9, 12.10 and 12.12) by standard analytical procedures. The results can be written in a conventional symbolic form as

$$\frac{d}{dt}(\overline{PE}) = -C(P, K),$$ (12.14)

$$\frac{d}{dt}(\overline{IE}) = \overline{Q} - C(I, K),$$ (12.15)

$$\frac{d}{dt}(\overline{KE}) = C(P, K) + C(I, K) - \overline{D}.$$ (12.16)

These equations are not so fierce as they look at first sight. The right-hand sides contain the global heating $\overline{Q}$, the global frictional dissipation $\overline{D}$ and two conversion terms. $C(P, K)$ is the rate of conversion of $\overline{PE}$ to $\overline{KE}$, corresponding to a lowering of the centre of gravity of the fluid as a result of net downward motion. $C(I, K)$ is the rate of conversion of $\overline{IE}$ to $\overline{KE}$ corresponding to the global effect of motion down pressure gradients. We shall not give the mathematical definitions of the conversion terms since the verbal descriptions are adequate for our present purpose. It will be noticed, however, that equation 12.15 contains no term corresponding to the total pressure work term in the parcel balance equation (see equation 12.12). Equations 12.14–12.16 are in fact derived under the assumption that the fluid is bounded by either rigid stationary surfaces or regions of zero pressure (as in the case of the notional upper boundary of the atmosphere). Under these conditions – which we shall assume without comment from now on – the global rate of working of the pressure field is zero. This important result makes the global energetics considerably simpler than the moving-element energetics: it reflects the fact that the total pressure work corresponds only to a redistribution of IE within the fluid.

All the terms in equations 12.14–12.16 would normally be specified in watts (W ≡ Joules per second, J s^{-1}). $\overline{Q}$, $C(P, K)$ and $C(I, K)$ are generally made up of positive and negative contributions from different parts of the fluid. The global frictional dissipation $\overline{D}$ is numerically equal to the global rate of working by friction, but has opposite sign. $\overline{D}$ is always non-negative.

Addition of equations 12.14–12.16 gives a relation for the rate of change of the global energy $\overline{E} = \overline{PE} + \overline{IE} + \overline{KE}$:

$$\frac{d\overline{E}}{dt} = \overline{Q} - \overline{D},$$ (12.17)

which is a particularly simple form because of the cancellation of the conversion terms. Figure 12.5 shows the box diagram representing equations 12.14–12.17.

When the left-hand sides of equations 12.14–12.16 are zero the global energetics are in a steady state. (In terms of figure 12.5 there is no net flow of energy into or out of any of the boxes.) In such a state

$$C(I, K) = \overline{D} = \overline{Q}$$ (12.18)

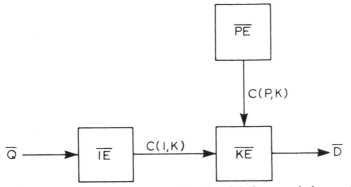

Fig. 12.5 Box diagram representing the global energy balance of a fluid bounded by either rigid stationary surfaces or regions of zero pressure. Each box represents the global amount of a form of energy – the amount obtained by summing the contributions from every fluid element in the system. When the global conversion rates $C(P, K)$ and $C(I, K)$ are positive, the accompanying arrows point in the directions shown. The directions of the arrows accompanying the global heating $\bar{Q}$ and the global frictional dissipation $\bar{D}$ correspond to both quantities being positive. $\bar{D}$ is in fact always positive, because the *global* effect of friction is always to oppose motion (subject to the assumed boundary conditions). $\bar{Q}$ could be negative in an unsteady state of the global system, but in a steady state, when it represents solely the frictional heating, $\bar{Q}$ must be positive.

and

$$C(P, K) = 0. \qquad (12.19)$$

Equations 12.18 and 12.19 represent the basic energy cycle of a fluid system whose global energetics is in a steady state. To a good first approximation, the Earth's atmosphere is a system of this type. (Note that steadiness of the global energetics does not imply steadiness locally: there may be all manner of fluctuations in local energy balances.)

What can be deduced from equations 12.18 and 12.19? Equation 12.19 says that the rate of conversion of $\overline{PE}$ to $\overline{KE}$ must be zero in the global steady state. This does not mean, of course, that there can be no vertical motion in the system, but merely that descent and ascent must balance one another in such a way that the centre of gravity of the fluid remains at the same height. Equations 12.18 say that the motions in the system must, in the globally steady state, convert IE to KE at the same global rate as KE is dissipated by friction and at the same rate as IE is produced by the global heating. Since frictional dissipation contributes identically to the heating, the global heating must consist entirely of the contribution of frictional dissipation, all other agencies of heating and cooling together giving a zero contribution. Equations 12.18 therefore express that $\overline{IE}$ and $\overline{KE}$ remain constant as a result of the global conversion $C(I, K)$ (by motion down pressure gradients) being balanced by the heating due to frictional dissipation of $\overline{KE}$.

Probably the only interesting and instructive feature of the basic energy cycle is that the global heating is positive: it is easy to deceive oneself into thinking that it should be zero. In any event, the basic energy cycle contains no explicit reference to the radiative heating and cooling, which are recognized to be the primary causes of the atmosphere's circulation (without which there would be no dissipation!). Later on, we shall discuss an energy cycle which contains *differential* heating as an essential part.

The hydrostatic global energy balance

The energy balance discussed in the previous section follows from the unapproximated equations of fluid motion. More frequently encountered, however, are energy balances for a fluid in which the pressure (p) and density (ρ) fields are assumed to be in hydrostatic balance:

$$\frac{\partial p}{\partial z} = -\rho g. \qquad (12.20)$$

Equation 12.20 has important repercussions for the energetics. Consider a column of air extending indefinitely upwards from ground level ($z = 0$). If the column has unit cross-section, the *PE* and *IE* within it at any time are given by

$$PE_{col} = \int_0^\infty \rho g z\, dz; \quad IE_{col} = \int_0^\infty \rho c_v T\, dz. \qquad (12.21)$$

If equation 12.20 holds, and the air can be treated as a perfect gas (so that $p = \rho R T$, where R is the gas constant for air), some straightforward algebraic manipulation of equation 12.21 shows that

$$PE_{col} = \frac{R}{c_v} IE_{col}. \qquad (12.22)$$

In a hydrostatic, perfect gas atmosphere the PE and IE contained in any column of air extending indefinitely upwards from z = 0 are in the ratio R/c_v. Under the stated assumptions a similar relation must exist between the global *PE* and *IE*.

Since R/c_v is about 2/5 for air, and departures from hydrostatic balance are generally small in atmospheric motions*, equation 12.22 is consistent with the observation that IE_1 and PE_1 are typically of the same order of magnitude, as shown earlier.

Because of the proportionality between PE_{col} and IE_{col}, it is usual to treat their sum as a single form of energy. This is usually called the total potential

---

* Excluding sound waves. For many other small scales of motion, the hydrostatic approximation is inappropriate for analyses in terms of vorticity, but the balance of vertical momentum is hydrostatic to a very good approximation.

energy (*TPE*), though it includes *IE* as well as *PE*! A further confusion is that the same term is generally used for the sum of the *global PE* and *IE*, presumably because of the undesirability of 'total total potential energy' or 'global total potential energy'. We shall refer to this quantity as $\overline{TPE}$.

In terms of $\overline{TPE}$, the hydrostatic global energy balance equations are

$$\frac{d}{dt}(\overline{TPE}) = \bar{Q} - C(T, K), \qquad (12.23)$$

$$\frac{d}{dt}(\overline{KE}) = C(T, K) - \bar{D}. \qquad (12.24)$$

$\bar{Q}$ has the same meaning as in the non-hydrostatic case (equations 12.14–12.16) but the other common terms have slightly different definitions. $\overline{KE}$ is now the global $\overline{KE}$ of horizontal motions only, and $\bar{D}$ is the global dissipation due to horizontal frictional forces only. $C(T, K)$ is the one conversion term: it represents the global rate of conversion of $\overline{TPE}$ to $\overline{KE}$ via motion down *horizontal* pressure gradients (whereas $C(I, K)$ in equations 12.15 and 12.16 includes conversion via motion down vertical pressure gradients). An alternative interpretation of $C(T, K)$ is useful. To a good approximation $C(T, K)$ represents the global effect of warmer air rising and cooler air sinking at the same level.

Figure 12.6 shows the box diagram which summarizes equations 12.23 and 12.24. The processes involved are heating and cooling, motion across horizontal pressure gradients, and frictional dissipation. In the steady state of the global system, equations 12.23 and 12.24 reduce to

$$C(T, K) = \bar{D} = \bar{Q}, \qquad (12.25)$$

which is the hydrostatic global energy cycle. It is obviously very similar to the non-hydrostatic energy cycle (equation 12.18), and, in particular, there is no explicit reference to non-frictional heating, or cooling. The key quantity in more sophisticated energy cycles is available potential energy, the general background to which we shall now discuss.

Available potential energy (*APE*)

Consider a room full of air, which, at some instant, is at rest and in hydrostatic balance. The state of rest will persist if (1) the surfaces of constant temperature and density are horizontal (so that there are no unbalanced pressure gradient forces) and (2) the temperature decreases upwards at less than the dry adiabatic lapse rate (so that the density stratification is stable) (see figure 12.7a). Under other conditions, however, the state of rest will not persist, although the initial $\overline{TPE}$ may be the same as in the case of continuing rest. For example (see figure 12.7b), if the surfaces of constant temperature and density are coincident but not horizontal, bulk flow will occur in response to the unbalanced horizontal pressure-gradient force and thus $\overline{TPE}$ will be converted to $\overline{KE}$. Again, if the surfaces of constant temperature and density are horizontal, but the temperature decreases upwards at more than the dry adiabatic lapse rate (see figure 12.7c),

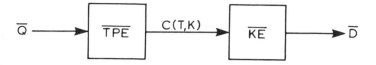

Fig. 12.6 Box diagram representing the hydrostatic global energy balance. $\overline{TPE}$ is the sum of $\overline{PE}$ and $\overline{IE}$, these two quantities being in a fixed proportion in an atmosphere which is in hydrostatic balance (see text). $C(T, K)$ is the global rate of conversion of $\overline{TPE}$ to $\overline{KE}$, representing the global effect of flow down horizontal pressure gradients.

any small disturbance may be expected to lead to overturning and therefore conversion of $\overline{TPE}$ to $\overline{KE}$.

From these simple examples we see that *the mere existence of IE and PE is no assurance that KE will be produced subsequently.* The determining factor is the spatial distribution of *IE* and *PE* rather than the global amounts. Also, it is

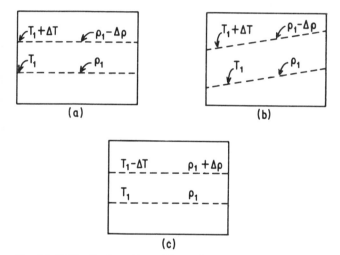

Fig. 12.7 Vertical sections through a room containing air with various distributions of temperature and density. The broken lines are lines of constant temperature and (to simplify matters) density also, in different states of instantaneous rest and hydrostatic balance. In (a) there are no horizontal pressure gradients and the temperature actually increases upwards: thus the system is in a state of (very) stable equilibrium. In (b) there is a horizontal gradient of hydrostatic pressure because the density increases from left to right (at all levels): motion will therefore occur, and (b) can be a state of only instantaneous rest. In (c), as in (a), there are no horizontal pressure gradients but the lapse rate of temperature is greater than the dry adiabatic (appreciably greater, because ρ increases upwards). Thus the system is in a state of unstable equilibrium, and any arbitrarily small disturbance will lead to bulk motion.

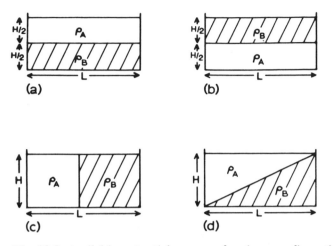

Fig. 12.8 Available potential energy of various configurations of equal volumes of two incompressible liquids. The density of an incompressible fluid is independent of pressure and so the global *IE* is unchanged in adiabatic processes. Thus calculation of the *APE* reduces to finding the reduction of global $\overline{PE}$ in the idealized rearrangement of mass; in other words, *PE* takes the place of $\overline{TPE}$ in calculating *APE*. In the arrangement shown in (a), a depth $H/2$ of liquid with density ρ_A overlies an equal depth of denser liquid (density $\rho_B > \rho_A$). Both liquids are imagined to be contained in a trough of width L and unit length (perpendicular to the plane of the paper). Taking $z = 0$ at the base of the trough, the centre of gravity of the lower liquid is at $z = \frac{1}{4}H$, and that of

clear that complete conversion of $\overline{TPE}$ to $\overline{KE}$ will not occur spontaneously, for such a change would require the air to have zero absolute temperature and to be concentrated at ground level ($z = 0$)! Typically only a small proportion of the $\overline{TPE}$ of a fluid system can be converted to $\overline{KE}$ by adiabatic processes.

In many applications, it is helpful to work in terms of that part of the $\overline{TPE}$ which could conceivably be converted to $\overline{KE}$ by adiabatic processes, rather than in terms of $\overline{TPE}$ itself. The convertible part of the $\overline{TPE}$ is called the *available potential energy* (*APE*). In practice, however, difficulties can arise in deciding to what extent conversion of $\overline{TPE}$ to $\overline{KE}$ is conceivable (especially in rotating systems) and alternative *definitions of APE* are frequently adopted. A useful one is the following: *the APE of a fluid system is its $\overline{TPE}$ less that of the state of minimum $\overline{TPE}$ obtainable from it by an idealized adiabatic rearrangement of the fluid**. We shall try to make this definition clearer by considering particular

---

* This is the definition used by Lorenz (1955, 1967). A different definition has been used by van Meighem (1973); see also Dutton and Johnson (1967). Pearce (1978) has proposed a definition which does not involve the state of minimum *TPE* and which is therefore essentially different from both Lorenz's and van Meighem's.

the upper liquid at $z = \frac{3}{4}H$, so the global PE is

$$\overline{PE}_a = \tfrac{1}{8}\rho_B g H^2 L + \tfrac{3}{8}\rho_A g H^2 L.$$

Since $\rho_B > \rho_A$, the configuration (a) is stable, and because the interface (and the upper free surface) is horizontal, any other arrangement of the fluids must have greater $\overline{PE}$ (so long as the boundary conditions imposed by the trough remain unchanged!). The APE is therefore zero and so (a) *is the reference state for the calculation of the APE of any other configuration.* In the case of (b), the denser liquid overlies the less dense, and the global PE is

$$\overline{PE}_b = \tfrac{1}{8}\rho_A g H^2 L + \tfrac{3}{8}\rho_B g H^2 L.$$

The APE of this arrangement is $\overline{PE}_b - \overline{PE}_a$, or

$$\overline{APE}_b = \tfrac{1}{4}(\rho_B - \rho_A)g H^2 L.$$

Similar reasoning applied to arrangements (c) and (d) shows that their APEs are respectively

$$APE_c = \tfrac{1}{8}(\rho_B - \rho_A)g H^2 L$$

$$APE_d = \tfrac{1}{24}(\rho_B - \rho_A)g H^2 L.$$

In each case the APE depends on the difference in densities. This reflects the fact that rearrangement to the reference state (a), while it involves lowering the centre of gravity of the denser fluid, also involves raising the centre of gravity of the less dense fluid.

APE calculations for compressible fluids are much more laborious but lead to broadly analogous results: differences in potential temperature are far more important than average values.

---

examples. The APE of the fluid configuration of figure 12.7b is simply its TPE minus that of the configuration obtained from it by rearranging the fluid adiabatically so that all the constant temperature and density surfaces are horizontal and the lapse rate is everywhere less than the dry adiabatic. The same goes for figure 12.7c's configuration. Quantitative examples give further insight: some cases involving incompressible liquids are considered in figures 12.8 and 12.9. It emerges that the APE depends on typical density or potential temperature *differences* in the fluid rather than on typical mean values of these quantities. The case dealt with in figure 12.9 is particularly important, because the initial state of stable stratification and a horizontal temperature gradient is similar to the usual condition of the troposphere. To a useful approximation (see figure 12.9), the APE of this configuration is proportional to the square of the horizontal temperature gradient and inversely proportional to the static stability. In the same approximation, therefore, APE is increased by heating fields which increase horizontal temperature gradients and/or decrease the static stability, and decreased by heating fields which have the opposite effects. Thus, APE is affected by *differential* heating rather than uniform heating.

On the basis of figure 12.9, the APE of the atmosphere may be expected to be only a small percentage of its *TPE*. If it is accepted that $\overline{KE}$ and *APE* are likely to be of the same order of magnitude, the observed smallness of typical values of KE_1 compared with typical values of IE_1 and PE_1 (shown earlier) is thus partly accounted for.

The hydrostatic global energy balance in terms of *APE*

In terms of *APE*, equations 12.23 and 12.24 become

$$\frac{d}{dt}(APE) = G(A) - C(A, K), \tag{12.26}$$

$$\frac{d}{dt}(\overline{KE}) = C(A, K) - \bar{D}. \tag{12.27}$$

$G(A)$ is the rate of generation of *APE*. Broadly, $G(A)$ is a measure of the extent to which the heating and cooling tend to increase existing horizontal temperature gradients and decrease existing static stabilities. (We shall invoke this interpretation several times in what follows. It is somewhat tedious to repeat

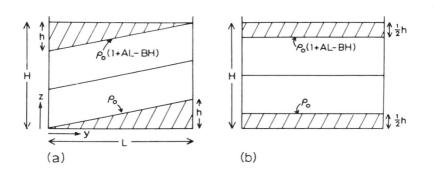

(a)   (b)

Fig. 12.9 Available potential energy of an incompressible fluid configuration with continuous horizontal and vertical variations of density. The two-fluid system examined in figure 12.8 is very easy to analyse, but a system with a continuous spatial variation of density is a better model for most geophysical applications. Diagram (a) shows a configuration in which the density varies with co-ordinates y and z (origin at lower left-hand corner of trough) as

$$\rho = \rho_0(1 + Ay - Bz).$$

Diagram (b) shows the state of minimum $\overline{PE}$ obtainable from configuration (a) by rearranging the fluid adiabatically. The *APE* of (a) is then the excess of the $\overline{PE}$ of (a) over that of (b). The required calculation need deal only with the change in *PE* of the shaded regions, since the rest of the fluid has its centre of gravity at $z = \frac{1}{2}H$ before and after the rearrangement, but the algebra is the whole interpretation so we shall mention only the part involving horizontal temperature gradients.) $C(A, K)$ is the same as $C(T, K)$ appearing in equation 12.24, and it therefore represents the conversion rate of $\overline{TPE}$ to $\overline{KE}$ due to motion down horizontal pressure gradients, or, approximately, the rising of warm air and the sinking of cold air at the same levels.

The box diagram representing equations 12.26 and 12.27 is shown in figure 12.10. In the steady state of the global system, equations 12.26 and 12.27 reduce to

$$G(A) = C(A, K) = \bar{D}, \qquad (12.28)$$

which is the global energy cycle in terms of *APE*. The rate of generation of *APE*, the rate of conversion to *KE*, and the rate of dissipation of *KE* by friction must be equal, in the steady state global mean. In the Earth's atmosphere $G(A)$ predominantly represents the tendency of low-latitude warming and high-latitude cooling to enhance the Pole–Equator temperature gradient. Equations 12.28 therefore directly involve the radiative heating and cooling process which are the ultimate cause of the atmospheric circulation. Furthermore (in the case of the atmosphere, at least), the contribution of frictional heating to $G(A)$ is very

---

nevertheless rather involved in the general case. Fortunately a limiting case can be done straightforwardly. If $h \ll H$ (so that, in (a), the difference in density $\Delta\rho_v$ between bottom and top is much greater than the difference $\Delta\rho_h$ between the two sides) the mass of the lower triangle of liquid is close to $\frac{1}{2}\rho_0 hL$ and its centre of gravity is close to $z = \frac{1}{3}h$. In (b) the same volume of fluid has its centre of gravity near $z = \frac{1}{4}h$, and the *decrease* in its *PE* is thus about

$$\Delta PE_- = \tfrac{1}{24}\rho_0 gh^2 L.$$

Similarly, the mass of the upper triangle of liquid is close to $\frac{1}{2}\rho_0(1 - BH)gh^2 L$, and the rearrangement from (a) to (b) leads to an *increase* of its *PE* of about

$$\Delta PE_+ = \tfrac{1}{24}\rho_0(1 - BH)gh^2 L.$$

The *APE* of (a) is therefore well approximated by

$$\Delta PE_- - \Delta PE_+ = \tfrac{1}{24}\rho_0 gh^2 BHL$$

and so

$$APE \simeq \frac{\rho_0 gBHL}{24} \left(\frac{AL}{B} \right)^2 = \rho_0 \frac{gH^2 L(\Delta\rho_h/\rho_0)^2}{24(\Delta\rho_v/\rho_0)}.$$

This turns out to be a reasonable approximation even if h is comparable with H (the error is only 20% when $h = H$). It can also be shown that a result of similar form arises when the model fluid is compressible. The major change is that $(\Delta\rho_h/\rho_0)$ and $(\Delta\rho_v/\rho_0)$ are replaced by the corresponding fractional variations of potential temperature.

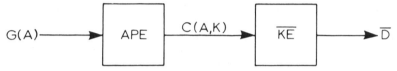

Fig. 12.10 Box diagram representing the global hydrostatic energy balance in terms of available potential energy (APE). $G(A)$ is the global rate of generation of APE, representing the extent to which the heating tends to enhance horizontal temperature gradients and reduce vertical stabilities. $C(A, K)$, the rate of conversion APE to $\overline{KE}$, is the same as $C(T, K)$ in Figure 12.6 and so represents the conversion to $\overline{KE}$ by motion down horizontal pressure gradients.

small indeed. In both respects, equations 12.28 constitute a far more useful and significant energy cycle than do equations 12.18 and 12.25.

The three quantities APE, $\overline{KE}$ and $G(A)$ (or $C(A, K)$, or $\overline{D}$) represent fundamental physical aspects of a fluid system which is in a global quasi-steady state. KE is a measure of typical (horizontal) flow speeds, APE gauges the extent to which the system departs from a state of uniform temperature and density in the horizontal and $G(A)$ is the rate at which energy traverses the system. Accounting for their values in the Earth's atmosphere, or in any similar fluid system, requires a comprehensive physical understanding and is an important objective of geophysical fluid dynamics.

By subdividing APE and KE into parts associated with different facets of the circulation, it is possible to construct more elaborate energy cycles which can be used to characterize the fluid system in further quantitative detail and also qualitatively. One such cycle is the zonal average/eddy cycle – the cycle which results when APE and KE are divided into the parts associated with the zonal (longitudinal) average fields and the parts associated with the eddy fields (the deviations from zonal averages). Figure 12.11 shows the box diagram representing the global energy balance in zonal average/eddy terms. Complete definition and explanation of the relevant quantities would involve a lengthy discussion, so we shall give approximate physical interpretations only.

The four subdivisions of energy are zonal average and eddy APE and KE, respectively A_Z, A_E, K_Z, K_E. A_Z represents the contribution APE from the zonal average temperature gradient between Pole and Equator; A_E is the remaining part of APE (so $A_E + A_Z = APE$), representing the contribution from temperature gradients in the east–west direction. K_Z represents the contribution to KE from the zonal average flow fields, and K_E is the remaining part of $\overline{KE}$ (so $K_E + K_Z = \overline{KE}$) representing the contribution from the eddy flow fields. A_Z, A_E, K_Z and K_E can be calculated from available observations and the values are frequently used to test numerical models of the atmosphere's circulation. A common fault of the models is that the mid-latitude cyclones are less intense than in the real atmosphere. This is reflected in the usual smallness of model-generated values of K_E compared with observed values.

The rate of generation of APE, $G(A)$, is divided into the rates of generation of

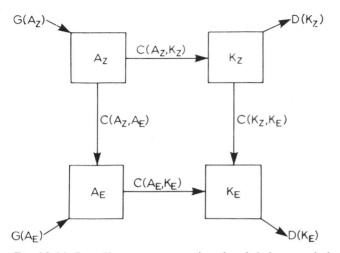

Fig. 12.11 Box diagram representing the global energy balance in terms of zonal average and eddy *APE* (A_Z and A_E) and $\overline{KE}$ (K_Z and K_E). The directions of the arrows correspond to positive values of the associated generation, dissipation and conversion rates ($D(K_Z)$ and $D(K_E)$ are necessarily positive). Diagrams of this sort are frequently constructed to represent the quasi-steady state energy cycle of a fluid system, according to observation or numerical simulation. In the case of the Earth's atmosphere all quantities are usually specified as the global values divided by the horizontal area covered — ideally the area of the entire Earth, but frequently only a hemisphere. Values of A_Z, K_Z, A_E and K_E are therefore quoted in J m^{-2}, and the generation, dissipation and conversion rates in W m^{-2}.

A_Z and A_E, respectively $G(A_Z)$ and $G(A_E)$. $G(A_Z)$ represents the extent to which the heating and cooling tends to enhance the zonal average temperature gradient between Pole and Equator, while $G(A_E)$ represents the extent to which it tends to enhance east—west temperature gradients. Similarly, the frictional dissipation rate D can be divided into the rates of dissipation of K_Z and K_E, $D(K_Z)$ and $D(K_E)$, corresponding to the effects of zonal average and eddy frictional forces.

The conversion rates $C(A_Z, K_Z)$ and $C(A_E, K_E)$ together constitute $C(A, K)$. $C(A_Z, K_Z)$ is the contribution from the zonal average motion — ascent in warm latitudes and descent in cool latitudes, warm and cool being defined relative to mid-latitude temperatures. $C(A_E, K_E)$ is the contribution from the eddy motion — ascent in warm longitudes and descent in cool longitudes, warm and cool being defined relative to the longitudinal average. There are two other conversion rates. $C(A_Z, A_E)$, the rate of conversion from A_Z to A_E, represents heat transfer from warm latitudes to cool latitudes by eddy motion. $C(K_Z, K_E)$, the rate of conversion from K_Z to K_E, represents eddy momentum transfer from latitudes of strong zonal (westerly) flow to latitudes of weaker zonal flow.

In a steady state of the global system, $G(A)$ and $C(A, K)$ must be positive, because D is always positive. The energy cycle embodied in equation 12.28 therefore allows only quantitative variation between different systems and external conditions. The zonal average/eddy cycle, because it has more brands of energy, is less constrained. Thus, the sum $C(A_Z, K_Z) + C(A_E, K_E)$ must be positive (so that $C(A, K)$ is positive) but energy arguments give no reason why they should both be positive. The same applies to $G(A_Z)$ and $G(A_E)$. Similarly, $C(A_Z, A_E)$ and $C(K_Z, K_E)$ may be of either sign. Obviously, there are many qualitatively different energy cycles according to different signs of $G(A_Z)$, $G(A_E)$ and the four conversion terms (note that $D(K_Z)$ and $D(K_E)$ must both be positive, however). Of course, some of these may not be realizable, for a process which is energetically possible may be forbidden by other dynamical constraints, but, on the other hand, it would be surprising if only one of them were realizable. Two examples are shown in figure 12.12. In the Hadley cycle the eddy components play no part and A_E and K_E are both zero; $C(A, K)$ consists entirely of the contribution of the zonal average flow. In a Rossby cycle, the eddy contribution dominates $C(A, K)$ and K_Z is maintained (against the dissipation $D(K_Z)$) by eddy momentum transfers giving a negative value of $C(K_Z, K_E)$. Several cycles of the Rossby type have been observed – the cycle of the Earth's atmosphere is one of these. The dynamical regime associated with the Hadley cycle is easily set up in laboratory systems.

Clearly, the zonal average/eddy cycle of a fluid system can give a qualitative characterization of that system. Quantitative values of the generation, dissipation

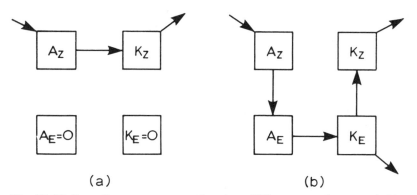

(a)   (b)

Fig. 12.12 Box diagrams representing two different zonal average/eddy energy cycles. The absence of an arrow indicates that the associated conversion rate is zero; the presence of an arrow indicates that the conversion occurs in the direction shown. (a) the Hadley cycle, (b) a Rossby cycle. The Rossby cycle shown in (b) has $C(A_Z, K_Z) = 0$ to emphasize the difference from the Hadley cycle and the importance of the negative value of $C(K_Z, K_E)$. In practice $C(A_Z, K_Z)$ may be non-zero and of either sign, but it is frequently a small term in the energy cycle: observations suggest that this is the case in the Earth's atmosphere. ($G(A_E)$ is also non-zero in the Earth's atmosphere).

and conversion rates are also important, though some of these are difficult to measure accurately. Both aspects are very useful for testing numerical models and are widely used for this purpose.

Many other energy cycles can be devised by defining the basic categories of *APE* and *KE* differently. Energy balances for limited regions can also be drawn up if boundary fluxes are accounted for. We leave the reader to savour such delights for himself — in almost any issue of the *Quarterly Journal of the Royal Meteorological Society, Journal of the Atmospheric Sciences*, or *Tellus*.

References

Atkinson, B. W. (1981) 'Weather, meteorology, physics, mathematics', this volume, 1—7.

Dutton, J. A. and Johnson, D. R. (1967) 'The theory of available potential energy and a variational approach to atmospheric energetics', *Adv. Geophys.*, 12, 333—436.

Lorenz, E. N. (1955) 'Available potential energy and the maintenance of the general circulation', *Tellus*, 7, 157—67.

Lorenz, E. N. (1967) *The Nature and Theory of the General Circulation of the Atmosphere*, Geneva, WMO.

Panofsky, H. A. (1981a) 'Atmospheric hydrodynamics', this volume, 8—20.

Panofsky, H. A. (1981b) 'Atmospheric thermodynamics', this volume, 21—32.

Pearce, R. P. (1978) 'On the concept of available potential energy', *Quart. J. R. Met. Soc.*, 104, 737—55.

van Meighem, J. (1973) *Atmospheric Energetics*, Oxford, Clarendon Press.

13

Trough-ridge systems as slant-wise convection

J. S. A. GREEN
Imperial College,
University of London

The continual succession of travelling cyclones (or troughs) and anticyclones (or ridges) in middle latitudes are important for weather forecasts because of the fluctuations in weather that they cause. But because systems of this size transfer heat from equatorial to polar regions, they also play a fundamental role in determining the global-scale temperature contrast. This in turn determines the mean winds and the global-scale climate. When we examine the mechanics of the eddies from this fundamental point of view we need concepts that are radically different from those familiar to a synoptic meteorologist. But, by using these different concepts, we are able to describe many aspects of the structure and energetics of the eddies. Such theory about the behaviour of these systems is rather important if we are to understand the processes that determine climate and its trends.

Cyclones as waves

During the week or so that such systems retain their identity, the upper air (travelling at say 30 m s^{-1}) will have moved some 20000 km – half way round the earth! – and so have gone through several such systems. In trying to describe this situation one must therefore think in terms of the global-scale assembly of isobaric patterns, rather than in terms of individual, isolated systems. Figure 13.1 shows that on this extended time scale the motion looks remarkably wave-like. Just as with waves on the surface of the ocean, not all waves will be in the same state of development at any one time. Some may be just beginning to develop, some may be performing the analogy of breaking and there may be groups of waves with larger amplitude and some groups of smaller amplitude. Figure 13.2 shows the kinetic energy of the wind at different scales, as discussed in the chapter on spectral analysis (Reiter, 1981). Notice that most of the kinetic energy is concentrated between wavelengths of 6000 and 3000 km. These wavelengths can be loosely identified with the distance between troughs as seen on synoptic charts of the middle latitudes. If the motion had been a pure unmodulated single wave, then the spectrum would have been in the form of a single line. From the

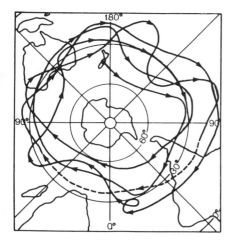

Fig. 13.1 Track of a constant-level balloon, flown at the 300 mb level in the Southern Hemisphere. Notice the wavelike nature of the track, with a rather well-defined wavelength of some 6000 km.

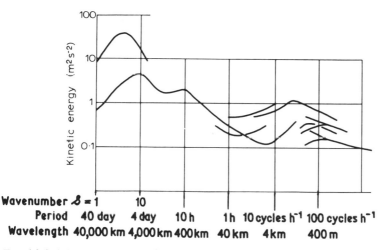

Fig. 13.2 Kinetic energy of the wind at different wavelengths. Each curve is from a different study. Notice the dominant peak in the wavelength range from 3000–6000 km which is the spectral representation of the trough–ridge scale of motion. A small bump near 500 km shows the frontal scale, that near 1 km the cumulus scale. Notice that both the kinetic energy and the wavelength are plotted on logarithmic scales.

peaked nature of the actual spectrum, one must conclude that the idea of such a simple wavelike form for the motion has at least some descriptive validity.

Properties of the waves

The wavelike pattern of velocity, pressure and temperature will, in general, move at a speed different to that of the air. This is a taxing concept, but we can see it in the simple example of waves on the free surface of a liquid, such as ocean waves. Here, it is clear that the surface undulation sweeps rapidly through the water (and past the piece of driftwood floating in it) carrying the velocity field. This causes the water (and the driftwood) to oscillate *much more slowly* than the wave speed, to and fro and up and down. Atmospheric waves are more complicated because, as we can infer from general observation: the pattern of velocities moves faster (about 10 m s^{-1} for the speed of movement of a typical trough) than the mean speed of the low-level air (about 0 m s^{-1}) but slower than the upper level air (about 30 m s^{-1} in a typical jet stream). As well as both the wavelength and the wave speed any complete theory must also describe the magnitude of the velocity (and temperature and pressure) perturbations carried by the wave.

Waves versus overturning

Because our interest is mainly in the energy transfer due to this scale of motion it is desirable to enquire why such a complicated mechanism is required. One could very well suppose, for example, that air warmed in tropical regions might ascend and move poleward while air cooled in polar regions would sink and move equatorward, thus transferring energy from tropics to pole. Indeed, this is exactly what would happen if the earth did not rotate and something like it appears to happen on Venus, which rotates very slowly. Consider in a little more detail, however, what would happen in a rotating atmosphere.

Heating in the tropics causes the columns of air there to expand, polar cooling leads to contraction so that, particularly at upper levels, the isobaric surfaces become tilted as in figure 13.3. Pressure is then higher on the equatorward side of a parcel of air (such as that marked A) than on the poleward side. Thus, that parcel and the whole ring of air round the globe will move polewards. But, since

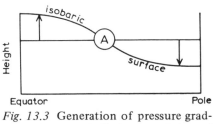

Fig. 13.3 Generation of pressure gradients by the latitudinal variation of heating.

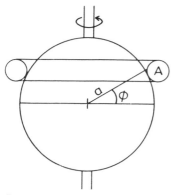

Fig. 13.4 Generation of zonal winds by latitudinal displacement of rings of air.

it is a complete ring of air (figure 13.4), it must conserve its angular momentum; so, as the radius of the ring decreases the air will begin to rotate faster. This in turn generates a centrifugal acceleration whose horizontal component tends to resist the pressure gradient. (The vertical component is negligible in comparison with the force of gravity.)

Some mathematical analysis is useful in order to determine the strength of this rotational effect. All that is needed is the law of conservation of angular momentum, but the zonal component of the momentum equation can be used instead. We have $\partial p/\partial x = 0$ because there is no torque – though this is not true to the extent that mountains may support a pressure difference between their upwind and downwind sides – so here we must strictly confine our attention to that part of the atmosphere above the level of such barriers. The momentum equation then reads $Du/Dt = fv$ but $v = Dy/Dt$ by definition so $Du/Dt = fDy/Dt$ and integrating, $u - fy =$ constant for each parcel of air, where f is the mean value of f over the range of y considered. In these equations, x and y are respectively W—E and N—S directions, u and v are the velocity components along those directions and f is the Coriolis parameter. Thus, changing the latitudinal distance of a parcel of air in a ring by distance δy changes its zonal component of velocity by $\delta u = f\delta y$. For example, for the air to acquire a zonal velocity of 10 m s^{-1} requires a displacement of only 100 km in middle latitudes where $f \simeq 10^{-4}$ s^{-1}. But such a zonal velocity is the magnitude of the wind that balances realistic horizontal gradients of pressure. We conclude that if the heating is realistic then a small meridional adjustment of the air will bring it into (geostrophic) balance with the pressure and that no further meridional displacement will occur.

Generalizing from this result we can say that a thermally driven meridional circulation is impossible if the virtual displacement δy is much less than the Pole—Equator distance, or if $u/fa \ll 1$, where a is the radius of the planet. For the earth the term u/fa (known as the Rossby number) has a value of about 0.02, showing that a simple meridional overturning is impossible. For Venus, the number is probably quite large showing that meridional overturning is present there.

Energy for the waves

Zonally symmetrical meridional overturning fails because angular momentum must be conserved, and thus acts to balance the driving force which is due to the distribution of temperature. How is this constraint to be avoided? If the air were to move poleward at one longitude and equatorward at another, the angular momentum could be shared between them by the action of the pressure field in the zonal direction. In this way, the fluid avoids the constraint imposed by the requirement that each parcel must keep its original angular momentum. Moreover, the motion envisaged has the wavelike character observed. It is not obvious that energy is available to drive the motion but we now show that gravitational potential energy is available.

As an initial state, suppose that the latitudinal variation of heating has made the tropics warmer than the poles and that local ordinary convection has made the vertical decrease of temperature less than the dry adiabatic: the potential temperature θ then *increases* with height and *decreases* with latitude as shown in figure 13.5.

Consider the change in the total gravitational potential energy of the system if parcel '1' were exchanged adiabatically with parcel '2' of figure 13.5: the potential energy resulting from this change will be available to provide kinetic energy to drive the exchange process. We see immediately from figure 13.5 that the parcel with the higher potential temperature θ_1 ascends and that with the lower potential temperature θ_2 descends. Notice that this is similar to the parcel theory of ordinary convection where potential energy is also released because the air ascending is warmer than the descending air. Thus, so far as the energetics

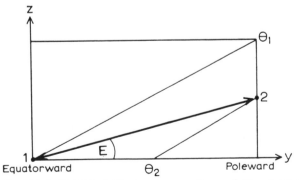

Fig. 13.5 Height–latitude cross-section of the troposphere showing potential temperature θ increasing equatorward and upwards. Even though the system is stable for vertical convection, energy can be released if particles are exchanged along the thick solid arrow.

is concerned the motion illustrated in figure 13.5 is convection, but such that the parcel paths are nearly horizontal. Because of this feature, and to distinguish the process from the motion that arises when the vertical lapse of temperature is unstable ('ordinary convection') the process is called 'slant-wise convection'. The term 'baroclinic' is also used to describe this motion. It is used in the sense that the isobaric and isothermal surfaces are inclined to each other, with the isobaric surfaces being very nearly horizontal (slope of say 10^{-4}). Unfortunately, the term is not a good one because all convection is baroclinic in this sense. What is more, motion in a statically stable atmosphere (where potential energy represents a restoring force for an oscillation) is also baroclinic.

A simple observable consequence of slant-wise convection

Is there evidence for this slant-wise orientation of the motion? It is difficult to check the proposed correlation between vertical and horizontal velocities directly because the vertical component of velocity is so small (about 1 cm s^{-1}) compared to turbulent fluctuations. However, anyone who has constructed a series of upper level charts (say 500 mb) will have become aware that the isotherms are not 'carried by the horizontal wind', i.e. advected, but move at about half that speed. If the motion were purely horizontal the isotherms would indeed be advected by the horizontal wind. But, as we have seen, the cold air descends and warms adiabatically, so that on a horizontal plane the isotherms appear to move less rapidly than the wind. If the particle paths have half the slope of the isentropic surfaces (surfaces of constant θ), it follows that the isotherms will move at half the speed of the wind. We show later that this is indeed the most efficient thermodynamic orientation.

Energetics

For an idealized process such as that shown in figure 13.5 the amount of energy released can be calculated. Calculation is not easy, but the results are of great importance for determining the scale and the intensity of the waves. As in the parcel theory of ordinary convection we imagine a situation in which parcel 1, when displaced adiabatically to its new level, just fits into the space vacated by parcel 2 and *vice versa*. Under this condition, the pressure field in the fluid is the same both before and after the exchange, and the potential energy released can only become kinetic energy of the motion (see White, 1981). If V_1 and V_2 are the respective volumes and p_1 and p_2 the pressures, the law of adiabatic expansion demands that

$$p_1 V_1{}^\gamma = p_2 V_2{}^\gamma, \qquad (13.1)$$

where $\gamma = c_p/c_v$ is the ratio of specific heats. The ratio of the masses of the two parcels follows:

$$\frac{M_2}{M_1} = \left(\frac{\rho_2 V_2}{\rho_1 V_1}\right) = \frac{\rho_2 p_1{}^{1/\gamma}}{\rho_1 p_2{}^{1/\gamma}} = \frac{\theta_1}{\theta_2}, \qquad (13.2)$$

where M and ρ are respectively mass and density.

We thus discover the importance of θ, 'the potential temperature', defined by $\theta = Cp^{1/\gamma}/\rho$, where C is a constant. Calculation of the potential energy released follows immediately. The change in potential energy (ΔPE) is given by the initial value minus the final value, as shown in equation 13.3.

$$\Delta PE = (M_1 g z_1 + M_2 g z_2) - (M_1 g z_2 + M_2 g z_1). \tag{13.3}$$

Algebraic manipulation gives

$$\Delta PE = M_1 g(z_1 - z_2) + M_2 g(z_2 - z_1) = M_1 g(z_2 - z_1)\left(\frac{M_2}{M_1} - 1\right) \tag{13.4}$$

$$= M_1 g(z_2 - z_1)(\theta_1 - \theta_2)/\theta_2.$$

From figure 13.5 we notice that z_2 is greater than z_1 and θ_1 is greater than θ_2, so that the term on the right-hand side of equation 13.4 is positive. This means that the final value of the PE is less than the initial value and thus the potential energy of the system is indeed decreased. Consider how this release of potential energy can be maximized with respect to the orientation of the motion. Suppose that the parcels were initially a distance, L, apart, then

$$z_2 - z_1 = L \sin E \quad \text{and} \quad y_2 - y_1 = L \cos E, \tag{13.5}$$

where E is shown in figure 13.5. The difference between their potential temperatures arises because of the horizontal gradient of potential temperature (say A) and the vertical gradient (say B) so that

$$\theta_2 - \theta_1 = (y_2 - y_1)A + (z_2 - z_1)B = (A \cos E + B \sin E)L, \tag{13.6}$$

thus,

$$\Delta PE = -(M_1 g L^2/\theta_2) \sin E(A \cos E + B \sin E). \tag{13.7}$$

Maximizing this expression with respect to variation in E gives

$$\frac{d}{dE}\{\sin E(A \cos E + B \sin E)\} = 0 \quad \text{or} \quad A \cos 2E + B \sin 2E = 0. \tag{13.8}$$

Thus, $\tan 2E = -A/B$ is the condition for optimum orientation of the particle paths.

Inferred properties of the large-scale waves

Considering now some values typical of the large scale: B is positive (statically stable) and $A/B \sim 2 \times 10^{-3}$ so $\tan E$ is small; thus $\tan E \simeq E$ and $E \simeq A/2B \sim 10^{-3}$. Consequently, the optimum slope of the particle paths is half the slope of the isentropic surfaces as anticipated from the temperature advection on 500 mb charts. The potential energy released is

$$\Delta PE \sim M_1 g L^2 A^2/4B\theta_2, \tag{13.9}$$

which is positive for all statically stable conditions.

This formula can be made even more quantitative and satisfying by considering the fluid velocities caused by the exchange shown in figure 13.5. Suppose each parcel acquires a typical velocity v then

$$\Delta PE = \tfrac{1}{2}M_1 v^2 + \tfrac{1}{2}M_2 v^2 = \tfrac{1}{2}M_1 \left(\frac{\theta_1 + \theta_2}{\theta_2} \right) v^2, \tag{13.10}$$

whence

$$v^2 = \frac{gL^2 A^2}{2B(\theta_1 + \theta_2)}. \tag{13.11}$$

Substituting typical values $LA \sim 10$ K, $\theta_1 + \theta_2 \sim 600$ K, $B \sim 5 \times 10^{-3}$ K m^{-1} gives $v \sim 14$ m s^{-1}. This is then the typical fluid velocity generated by a wave that can do the interchange shown in figure 13.5. In practice, the air is constrained to move nearly horizontally at both the ground and the tropopause, so in those areas it cannot release potential energy. Yet the air in those areas will share the potential energy liberated by less constrained parcels. This effect can be taken into account by calculating an optimum rearrangement for the whole atmosphere which gives an answer one third of the maximum value of release of potential energy or $v \sim 14$ m s$^{-1} \div \sqrt{3} \sim 10$ m s^{-1}, which is more typical of observed eddy velocities. We also infer a value for the vertical component of velocity $w \sim Ev \sim 1$ cm s^{-1}, which is consistent with observed rainfall rates.

A rather neat relation follows if we use, instead of the horizontal gradient of temperature, its value as given by the 'thermal wind equation':

$$\frac{\partial u}{\partial z} = -\frac{gA}{f\bar{\theta}}, \tag{13.12}$$

where $\bar{\theta} = (\theta_1 + \theta_2)/2$, which gives

$$v^2 = \frac{L^2 f^2 (\partial u/\partial z)^2 \bar{\theta}}{4gB}. \tag{13.13}$$

The value for v is as before, but now note that L/v is a typical time scale for the development (it is of the order of a few days) and that fL/v is a measure of the ratio, Coriolis acceleration to relative acceleration of a parcel (another Rossby number). Its value,

$$\frac{fL}{v} = 2 \left(\frac{g}{\theta} \frac{\partial \theta}{\partial z} \right)^{1/2} \left(\frac{\partial u}{\partial z} \right)^{-1} \tag{13.14}$$

shows that the Coriolis accelerations are about one order of magnitude greater than the inertial accelerations — in other words, that the horizontal flow will be nearly geostrophic — in spite of the vertical velocity being non-geostrophic and very important.

Another interesting feature of our formulation is that it also predicts something about the horizontal flux of the heat carried by the waves. The meridional component of fluid velocity is proportional to v and therefore to LA and therefore to the horizontal variation of temperature (say $\Delta\theta$). The temperature difference between the two parcels is also proportional to $\Delta\theta$ so the heat flux, which varies as the product $v\Delta\theta$, is proportional to $(\Delta\theta)^2$. Now, in the northern hemisphere winter, the temperature contrast $\Delta\theta$ is about twice the summer value and observation closely confirms the prediction that the energy flux should vary by a factor of about 2^2 between these two seasons.

Wavelength of the large-scale eddies

In the above analysis, we have tacitly admitted the existence of a wavelength because of the different longitudes at which the two parcels travel in their opposite directions; but there is no direct information as to their likely separation. A consideration of the time scale L/v provides a clue. This must be the same time scale as for parcels travelling through the wave. The zonal component of the wind is sheared in order to balance the latitudinal temperature gradient and we can suppose tentatively that the wave speed is roughly the mean flow speed say $\frac{1}{2}H(\partial u/\partial z)$ where H is the depth of the troposphere. If the wave has length l, then the periodic time for a parcel passing through it is $l/\frac{1}{2}H(\partial u/\partial z)$ and the corresponding radian measure (l/frequency) is $l/\pi H(\partial u/\partial z)$. Equating this to the time scale L/v gives an expression for the wavelength:

$$l \simeq \frac{2\pi H}{f}\left(\frac{g}{\theta}\frac{\partial\theta}{\partial z}\right)^{1/2} \tag{13.15}$$

Surprisingly, there is no reference to the shear $\partial u/\partial z$ in this formula, but only to H, the depth of the sheared layer, and to the proportionality factor $\frac{1}{f}\left(\frac{g}{\theta}\frac{\partial\theta}{\partial z}\right)^{1/2}$
For tropospheric values this is about 100.

Long waves and secondary cyclones

For tropospheric motion H is 10–15 km and equation 13.15 gives values of l ranging from 6000–9000 km, in accord with the longer waves shown in figure 13.1. In the frontal zones generated by these long waves the shear and horizontal temperature gradient are typically compressed until they extend upwards 1 or 2 km, and these zones are associated with secondary cyclones whose length is 600–1000 km for values of H ranging from 1–2 km, as is often observed. Note that secondary cyclones are sometimes called wave cyclones because their isobars often appear less closed than in the larger systems. Since the closed nature of the isobars indicates little about the fluid motion, and all cyclone systems can be represented as waves we believe this nomenclature to be misleading.

Weather

These smaller-scale motions are energetically rather trivial except that, just because the air converges and diverges over very much shorter horizontal scales than for the longer waves, the vertical velocities in them are larger than those in the long waves. Thus, almost all are accompanied by very heavy rainfall (or, as with polar lows, snowfall). Again, even fairly casual observation verifies that heavy, unexpected rain is often associated with a secondary trough that is barely detectable on a conventional analysis. This poses problems for numerical weather prediction because the resolution is usually chosen so as to resolve the longer waves and the shorter, more weather-intensive waves are poorly represented. A remedy might be to use a three-layer type of model in which the lower two layers were about one kilometre apart, so as to resolve the short shallow waves, with the third layer near the tropopause to represent the long waves. So far as I know this has never been tried as a scheme of numerical weather prediction.

Constraints on airflow

In the first part of this chapter we were able to deduce many observable properties of actual systems using only arguments about their energetics. We did not show that the exchange anticipated is consistent with the constraints represented by the equations of motion of the air. The most powerful constraints are imposed by the observation that the time scale of the motion is rather long compared to a pendulum day so that the motion must satisfy (1) the geostrophic wind, (2) the thermal wind, (3) a vorticity equation which states that parcels of air must conserve their vorticity (local spin) except that vertical velocities cause cyclonic vorticity by stretching the vortex lines of the Earth's solid rotation (see Harwood, 1981). This relation can be written:

$$\frac{D\zeta}{Dt} \simeq f \frac{\partial w}{\partial z} \qquad (13.16)$$

where ζ is the vorticity, D/Dt means rate of change of a property of a parcel of air (the substantial derivative), w is vertical velocity and z height. Let us construct a graphical solution that satisfies these equations.

Consider a vertical section running west—east through a long wave (figure 13.6). We have a pressure field with (say) a minimum at p_- and maxima at p_+. We immediately infer equatorward movement of air in the region marked v_- and poleward movement in the region marked v_+ because of geostrophic balance. It follows that the air will be warm (θ_+) in this section near v_+ and cool (θ_-) near v_- because of the horizontal advection of temperature even allowing for some adiabatic warming and cooling due to the slantwise paths. Hydrostatic balance (or, equivalently, the thermal wind equation) requires that the trough and ridge lines should slope towards the west with increasing height. Note that this slope is obvious on synoptic charts showing amplifying waves: it is a valuable rule for deciding if a wave will amplify further or not.

Now we would like to engineer the system so that the warm air can go upward (w_+) and the cold air downward (w_-). Here we meet the final constraint imposed by the vorticity equation. The line p_- represents the trough line, or line of cyclonic vorticity where the air must have positive or cyclonic vorticity. But the vorticity equation states that air can acquire such vorticity only where $\partial w/\partial z$ is

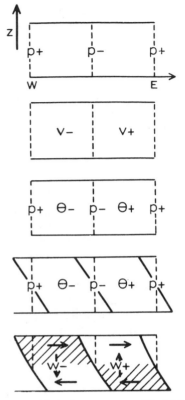

Fig. 13.6 A sequence of height–longitude sections through a wave showing how geostrophic, thermal wind and vorticity equations can be satisfied so long as the wave moves with the mean zonal velocity for the layer.

positive which is in the two shaded regions. Thus, air must arrive at the line p_- by moving from west to east at upper levels, but east to west at lower levels. In the unshaded regions $\partial w/\partial z$ is negative, and the air is gaining negative or anticyclonic vorticity, which is exactly what it needs to reach the anticyclonic region of high pressure represented by the line p_+. Notice that the diagrams that we have constructed are all *relative to the wave* and we see that we have just deduced the property mentioned in the introduction: the wave is moving faster than the air at low levels, so the relative motion of the air is east to west; and the upper air overtakes the wave, thus moving, relative to the wave, from west to east. Thus, our cyclone model is consistent with the constraints imposed by the equations of motion so long as the wave is 'steered' by some middle level flow.

Maturity and decay

As the wave grows in intensity, friction with the ground causes air to blow across the isobars towards low pressure (note the change of wind with height on a day with small cumulus where the surface wind is distinctly to the left (in the northern hemisphere) of the more geostrophic track of the cumulus clouds). This convergence induces upward motion in the trough tending to cool the air (the motion path there is now *steeper* than the isentropic surfaces) and the cold air slowly advances into the trough, while the opposite tendency occurs in the ridge. This is why the warm air advances away from the cyclone centre as it develops. Finally, the cyclonic area is filled with cold air. The hydrostatic equation then demands that the trough is more intense at upper levels than at lower levels and we have an occluded cyclone. Moreover, because the intensity is reduced at low levels, surface friction is not as powerful a brake and this state of affairs can persist for a long time unchanged — the typical 'old cold low' that may persist for a week or more. Again, by this time, the air in the low will have been lifted quite a long way so there may be substantial layer cloud and gentle precipitation.

Long waves and β

The nature of large wavelength waves is a response, not only to the rotation of the earth, but also to its difference at different latitudes. Air coming from equatorial regions has a deficit of vorticity, air from polar regions an excess. Figure 13.7 illustrates qualitatively what will happen. Air to the east of the trough arrives with less (i.e. equatorial) vorticity than expected, while the air to the west of the trough arrives with more (i.e. polar) vorticity. Thus, the vorticity pattern and therefore the wave tends to move towards the west compared to the situation without the variation of Coriolis parameter. The simplest theory (developed by Rossby) shows that the retardation is $\beta l^2/4\pi^2$ where $\beta = \partial f/\partial y$, which amounts to a retardation of some $5-10$ m s^{-1} for the long waves. Thus, these waves move detectably slower than the mean tropospheric flow. For various reasons connected with oversimplifications of the problem, this value for the retardation is something of an overestimate and retardations of $3-7$ m s^{-1} are more realistic. Nevertheless, somewhat longer waves will suffer a retardation

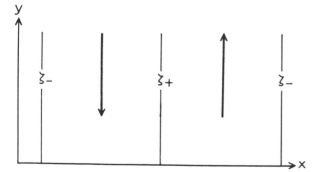

Fig. 13.7 Retardation of wave-speed due to variation of Coriolis parameter with latitude.

comparable with their phase speed. We begin to recognize in these waves some features of the blocking situations in which the global circulation seizes up because of resonance of these long nearly-stationary waves of slant-wise convection with topographic forcing.

Transfer properties: displacements

Air travelling rapidly relative to a wave (say at the ground and tropopause) spends little time under the influence of the north—south component of velocity whose pattern is carried at the wave speed. Conversely, at the level where the air moves at the same speed as the wave (the steering level) air will suffer considerable displacement because it is unable to escape from the wave. However, as the wave grows in amplitude, air on the equatorward side of the trough also begins to move in the same direction as the wave so that too suffers particularly great displacements towards the pole. Thus, we find a developing 'warm sector' filled with warm air of fairly uniform temperature that is unable to escape from the wave. Being warm air from low latitudes it is also quite humid so a large, fairly uniform area of precipitation develops.

Transfer of vorticity due to the waves

The vorticity transfer of these waves is complicated but important. It is significantly not what one would expect at first sight and its thorough study leads to an explanation of the main desert zones of the earth.

Our original consideration was to seek a mechanism for energy transfer that could avoid the penalty imposed by conserving angular momentum, but this has been replaced by the constraint of satisfying the vorticity equation. Consider the implications by reference to figure 13.8 which shows a latitude—height section

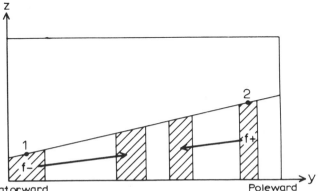

Fig. 13.8 Vorticity transfer as a result of the variation of Coriolis parameter and vortex-stretching.

through the wave. The line 1—2 represents the dominant parcel-displacement which we expect to find near the steering level. Parcels travelling from 1 towards 2 carry small planetary vorticity poleward while those travelling from 2 towards 1 (at a different longitude) carry large planetary vorticity towards the equator — an equatorward flux of cyclonic vorticity. In this respect, the motion tends to mix the planetary vorticity making the vorticity of the mean flow less in high latitudes and greater in low latitudes. The final state of such a mixing process would be for the air in the hemisphere to rotate with a constant mean Coriolis parameter which would give easterly winds in low latitudes and westerly winds in high. Friction with the underlying surface would reduce the enormous wind speeds otherwise deduced (about 200 m s^{-1}) but the picture would remain qualitatively as we have described it. Observation shows that the surface wind has a belt of westerlies only in middle latitudes so some other process must also be contributing.

The vorticity equation says that vortex-stretching is also important. But the parcel extending between the line 1—2 and the ground is being stretched ($\partial w/\partial z > 0$) as it moves from 1 towards 2 so its vorticity is anomalously cyclonic, whereas that moving from 2 towards 1 is being compressed ($\partial w/\partial z < 0$) so its vorticity is anomalously anticyclonic. This effect therefore tends to produce a poleward flux of *cyclonic* vorticity particularly in middle latitudes where the slope of the isentropic surfaces, hence the slope of the particle paths, is greatest. Notice that the opposite effect occurs in the region above the steering level, but that this is less effective than that below, since the steering level is nearer the lower boundary (because of the Rossby retardation) so the stretching there is greater.

Figure 13.9 shows qualitatively the contribution to the vorticity flux due to these two processes. The eddy motion we have been considering is confined to a broad belt in middle latitudes so the mixing of planetary vorticity is confined to the same broad latitude belt. The region of strong horizontal gradient of tempera-ture is confined to a smaller (about $30°$ to $60°$) latitude belt where the vortex

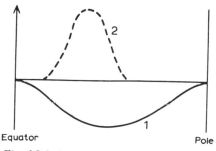

Equator   Pole

Fig. 13.9 Poleward flux of cyclonic
vorticity due to
   (1) mixing of planetary vorticity,
   (2) advection of vorticity-anomalies
        due to stretching of vortex.

Dynamical Meteorology

stretching effect is largest. If surface friction is not to exert continuously a torque on the air in the hemisphere (and that is the only other process that can change the vorticity), curve 2 must dominate curve 1 somewhere and we deduce that their sum must be as shown in figure 13.10, where a rough scale of latitude has been indicated to help description.

Between latitudes 0° and 15° there is an accumulation of cyclonic vorticity (shown in figure 13.10 as ζ). Between 15° and 45° a divergence of cyclonic vorticity exists: between 45° and 70° convergence and between 70° and 90° divergence (but small there because the eddy is not so intense).

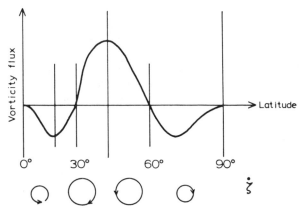

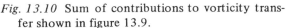

Fig. 13.10 Sum of contributions to vorticity transfer shown in figure 13.9.

Consequent surface wind and momentum transfer

The only way these imports and exports of vorticity can be balanced is by frictional interaction with the ground, and it is interesting to construct a surface wind field from the vorticity transfer. Eddy intensity is small in equatorial regions so the mean surface zonal wind vanishes there. Between 0° and 15° latitude the zonal wind must become easterly, because cyclonic vorticity is being injected into it by the eddies. Similarly, for the other latitude belts (figure 13.11). This picture is very much like the observed surface wind though there are geometrical distortions due to our qualitative treatment of the geometry of the earth and of the eddies.

The most curious aspect of the description emerges when we decide to interpret the action of these large-scale eddies in terms of their momentum transfer. We choose to infer this from (say) figure 13.11 and see that the waves must be injecting westerly momentum into the latitude belt 30° to 60° in order to support the surface westerly wind against friction. This momentum has been taken out of the air in low and high latitudes and the cycle is completed because westerly (positive) momentum is put back by friction acting on the (negative) easterly winds. But what a curious process this is, for it is exactly in the 30° to 60° lati-

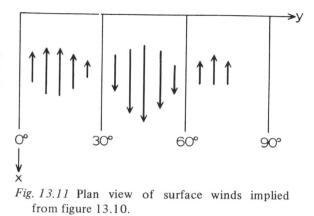

Fig. 13.11 Plan view of surface winds implied
from figure 13.10.

tude belt that the winds in the troposphere are most strongly westerly, and that is just where the waves are putting in more westerly momentum!

The uncommitted scientist is entitled to be sceptical about this. Perhaps it violates one's concept of the law of increasing disorder in natural systems? In fact it does not, because the energy of the system is almost entirely potential and as we have discussed in detail the potential energy is almost everywhere heavily and efficiently diminished. Momentum transfer is an incidental feature of the maximum efficiency of that more energetically important process.

Observation

Consider the structure of a wave that has this property of momentum transfer. As in the plan view of figure 13.12 we must imagine poleward-moving air to have an excess of westerly momentum compared to equatorward-moving air at least between latitudes $0°$ and $45°$. This gives the wave the asymmetric structure shown schematically in the figure which is indeed characteristic of and observable in the motion at (say) 500 mb, where the heights of the pressure surface show streamlines of the flow. Notice the characteristic strength of the south-westerly flow compared to the north-westerly. It really is very easy to *see* the air entering the tropics with little zonal momentum then going out with much more, while the explanation of *why* the wave has this structure is so complex.

In higher latitudes, the wave becomes symmetrical, showing no net transfer of momentum. It then becomes oriented in the opposite sense but the effect is so subtle that it can be lost in irregularities of analysis.

Orientation of the trough line

Skewness of the wave is associated with a tilting of the trough line between west of south to east of north (shown dashed in figure 13.12) and we might try to describe this by supposing that the stronger westerly wind of middle latitudes blows the wave along more rapidly there, hence giving the bowed shape which,

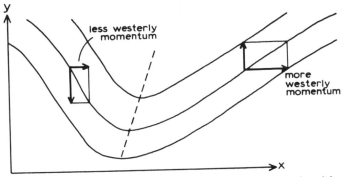

Fig. 13.12 Plan view of wave that transfers westerly (positive) momentum pole-ward. Notice that the north–south mass flux balances but the north–south flux of momentum is positive. A consequence of this is that the trough line (shown dashed) is not oriented north–south.

in turn, leads to the inferred momentum transfer. This is, at best, a half truth because the waves are observed to develop with this orientation, which changes little through their lifetime, whereas the latitudinal velocity differential would carry the slope very large distances (about 10 000 km in one week of lifetime). It is in fact characteristic of fluid systems that they tend to select some optimum shape for the pattern of motion which persists while the intensity of the motion changes greatly.

Implication of surface wind distribution

Explanation of the surface wind is lengthy and complex, whereas explanation of the upper wind – which is much stronger – demands only the application of the thermal wind relation. Now consider why the surface winds are so important. If there were no surface friction the surface winds would be in geostrophic equili-brium, with higher pressure near 30° latitude (the subtropical high), and lower near 60°. Surface friction accompanied by turbulence serves to slow down the wind in a layer near the surface (usually about 1 km deep) so that the wind is no longer strong enough to balance the pressure field. It moves across the isobars towards lower pressure (figure 13.13). This forms a new balanced state in which the air in the boundary layer gains energy by flowing towards low pressure, but loses it again through friction. Following the argument through for the surface easterlies gives equatorward flow as shown in figure 13.13. Continuity of mass then demands ascent in equatorial regions (the intertropical convergence) and near latitude 60°, and descent near 30° and 90°. Since the high-level air is relatively dry, deserts are formed under the descending branches of the circulation. Thus the transport of vorticity by long waves induces the great belts of deserts near latitude 30°.

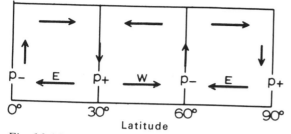

Fig. 13.13 Connection between surface winds and
the meridional circulation.

References

Harwood, R. S. (1981) 'Atmospheric vorticity and divergence', this volume, 33–54.

Reiter, E. R. (1981) 'The spectrum of atmospheric motions', this volume, 130–137.

White, A. A. (1981) 'Atmospheric energetics', this volume, 153–175.

14
Numerical modelling of the atmosphere

A. J. GADD
Meteorological Office,
Bracknell

During the past 30 years or so numerical modelling, facilitated by the development of ever more powerful electronic computers, has come to occupy a major role within dynamical meteorology. There are many aspects to numerical modelling; here we shall concentrate mainly on those which are concerned with the application of numerical methods to short-range weather forecasting.

The basic equations

We begin by considering again the seven atmospheric variables and the seven equations which describe their behaviour (see Panofsky, 1981a, b). The seven variables are the three components of the wind velocity vector (u, v, and w); the air density (ρ), temperature (T) and pressure (p); and the humidity mixing ratio (q). Other forms of the variables are sometimes used (e.g. potential temperature instead of temperature, see Panofsky, 1981b), whilst in some numerical models additional variables (e.g. liquid water content) are required. In this paper we limit ourselves to the seven variables listed above.

The seven equations which govern the behaviour of these seven basic variables have been described by Panofsky (1981a, b) in earlier chapters. They are the three equations of motion (one corresponding to each velocity component), the continuity equation (describing the conservation of mass), the thermodynamic or heat equation, the moisture equation and the equation of state. Along with the appropriate boundary conditions (e.g. at the earth's surface) these seven equations provide a mathematical description of the atmosphere.

Strictly speaking, of course, the equations provide an *approximate* description of the atmosphere, since we always introduce some simplifications. For example, we almost always use a single value for the acceleration due to gravity (g) and neglect the small variations in g from one place to another. Thus, in practice, the whole of dynamical meteorology is concerned with *model* atmospheres, in which the meteorologist's life has been made more bearable by the adoption of various simplifications and idealizations.

In particular applications, it may be possible to make further approximations and thus obtain simpler models. The resulting simplifications may be straightforward, as when the hydrostatic approximation (Panofsky, 1981a) is used for the vertical equation of motion. In other cases, the simplification may involve considerable manipulation of the equations and the use of derived variables such as vorticity (Harwood, 1981). Such approximations may have the result that the number of variables and equations used in the model is reduced from seven.

Solutions

We have referred above to the governing equations as a *description* of the behaviour of the atmosphere, but of course this description can be useful to us only if we are able to obtain *solutions* of the equations. Some problems in physics are governed by equations which, together with boundary or initial conditions, may be solved with pencil and paper by the application of mathematical techniques. The motion of a simple pendulum displaced from equilibrium is a familiar example. Solutions of this kind are technically known as 'analytical' solutions. Unfortunately, the equations of meteorology are not like this; they have no known analytical solutions. They are, in fact, rather difficult equations to handle; more difficult for example than the equations describing the motion of a spaceship towards the moon. (This is partly why weather forecasting seems to many people to be less successful than space travel.) Thus, for solutions of the meteorological equations we turn to numerical methods as outlined in this chapter.

It should be mentioned that analytical solutions have been obtained for simplified approximate forms of the meteorological equations (the brilliant work of Rossby and Eady was largely of this kind: see Atkinson, 1981a, b) and the analytical approach continues to be used to good effect by theoretical meteorologists to increase our understanding of atmospheric motions. But where we desire to predict a particular rather than a general situation in the atmosphere, or to take account of all the terms in our governing equations, we have no alternative to numerical modelling.

Choice of co-ordinates

In this chapter we shall for convenience talk about the meteorological equations in terms of familiar x, y, and z co-ordinates. This is quite adequate for the description of events at a single point on the earth's surface, where we may consider that the x axis points eastward, the y axis northward, and the z axis vertically upwards. It is important to bear in mind, however, that most numerical models of the atmosphere are obliged to adopt more sophisticated co-ordinate systems. In order to take simultaneous account of an extensive portion of the atmosphere, the earth's roughly spherical shape demands that we use either spherical polar co-ordinates or co-ordinates on a map projection, usually polar stereographic (figure 14.1) or Mercator. Associated with such co-ordinates are minor complications in the governing equations which we shall not consider here. In the vertical, pressure p is often used in place of altitude z as the co-ordinate (correspondingly

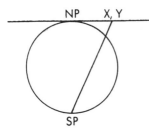

Fig. 14.1 Polar stereographic projection in which a point on the Earth's surface is projected from the South Pole (SP) onto the tangent plane at the North Pole (NP). The horizontal co-ordinates (x, y) are measured on this plane. These co-ordinates can be used for regional or hemispheric domains, but not for global models.

z replaces p as a variable of the model) if this allows some simplification of the equations. The complications associated with the irregular mountainous surface of the earth lead many modellers to adopt σ (pressure divided by the pressure at the earth's surface) as the vertical co-ordinate (figure 14.2). Once again, we merely note this complication and pass on to further general aspects of numerical models.

Grid points

The concept of grid points is fundamental to almost every numerical model of the atmosphere. (In this chapter we shall be considering only what are called finite difference models, in which grid points are used very directly. Even in alternative kinds of numerical model however, such as spectral models or finite element models, grid points are also in practice used to some extent.) The principle underlying the use of grid points is that a continuous fluid (the atmosphere) may be represented by values of the seven basic variables at a finite number of locations arranged in a three-dimensional array. Each of these locations corresponds to fixed values of the spatial co-ordinates (e.g. of x, y and z) and is referred to as

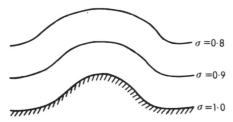

Fig. 14.2 Sigma co-ordinate system in which $\sigma = p/p_s$ is the vertical co-ordinate, p_s being the pressure at the Earth's surface. The Earth's surface thus coincides with the co-ordinate surface $\sigma = 1$ and the other co-ordinate surfaces follow the terrain.

a grid point. Notice that an element of approximation is immediately involved when we choose a set of grid points, since however close together our grid points are, there will always be motions of the atmosphere on a scale small enough that they slip through our mesh undetected (Pedder, 1981a, b, c). Thus, the spacing of the grid points, known as the grid length, is an important parameter. The grid length usually has to be chosen as a compromise between a desire for accuracy and the limitations of computer resources. Often there are different grid lengths associated with horizontal and vertical co-ordinates.

Suppose a set of grid points is chosen to describe some region of the atmosphere such that there are M grid points in each horizontal level and N levels in the vertical. Then, remembering that we have seven basic variables, we see that a collection of $7MN$ numbers defines the state of our model atmosphere at an instant of time, hence the term *numerical* model. The task of numerical modelling is usually, given an initial set of $7MN$ numbers, to calculate how each of them will change during some period of time; in other words to predict the evolution in time of the model atmosphere from some initial state.

Predictive and diagnostic equations

Our set of seven governing equations may be divided into two groups in a way which is very important from a numerical modelling point of view. The distinction depends on whether or not the rate of change with time of one of the seven variables appears in the equation. If an equation contains a rate of change with time it is called a predictive equation, otherwise it is called a diagnostic equation. For example, the equation of state, with the effects of water vapour included, is

$$p = R\rho T(1 + \epsilon q), \tag{14.1}$$

where R is the gas constant for dry air and ϵ is the ratio of molecular weights of water vapour and dry air. Given values of ρ, T and q we may use the above equation to diagnose the value of p. In contrast, the equation of motion in the x direction (see Panofsky 1981a) is

$$Acc_x = PGF_x + CF_x + Fr_x, \tag{14.2}$$

where *Acc*, *PGF*, *CF* and *Fr* represent acceleration, pressure-gradient force, Coriolis force and frictional force respectively and the subscript x indicates their components in the x direction. Now acceleration is rate of change of velocity, and therefore Acc_x is the rate of change of the variable u. Thus equation 14.2 is a predictive equation. Some equations may be either predictive or diagnostic depending on the particular approximations adopted in a given numerical model. In a similar fashion, the seven basic variables may, in a given numerical model, be classified as predicted or diagnosed. Notice that in order to specify the exact state of our model atmosphere we need only supply the grid point values of the predicted variables, since from these the corresponding values of the diagnosed variables may be calculated.

Finite differences

Apart from the equation of state, all the governing equations, predictive and diagnostic, contain spatial derivatives (i.e. rates of change with distance) of the basic variables. For example in equation 14.2 the pressure-gradient force is

$$PGF_x = -\frac{1}{\rho}\frac{\partial p}{\partial x}, \tag{14.3}$$

this being equation 2.4 of Panofsky (1981a).

In a widely used class of numerical models known as finite difference models, each spatial derivative is replaced by a finite-difference approximation. Suppose for instance that we wish to replace $\partial p/\partial x$ with a finite-difference approximation. Consider a small portion of a grid-point array (figure 14.3) at a particular horizontal level of a model atmosphere. The grid points are identified by a pair of indices (i, j) indicating their relative position, and the grid length is a. The simplest finite-difference approximation to the derivative $\partial p/\partial x$ at grid point (i, j) is the following

$$\left(\frac{\partial p}{\partial x}\right)_{i, j} = \frac{p_{i+1,j} - p_{i-1,j}}{2a}. \tag{14.4}$$

For obvious reasons this is known as a centred difference. If we recall our basic calculus we notice that the *definition* of the derivative is the limit of the right-hand side as a tends to zero. Thus, whereas our differential equations apply in the limit, we replace them by *difference equations* which stop short of the limit, with, of course, some implied loss of accuracy. The difference between the true value of a derivative and its finite-difference approximation is called the truncation error. This error can be reduced by using a more complicated finite difference

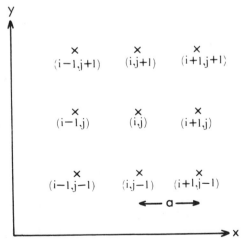

Fig. 14.3 A grid-point array for one horizontal level in the atmosphere.

in place of the simple centred difference; this complication is sometimes worthwhile, but nevertheless the simple form is very widely used.

It is worth noting that sometimes a numerical model uses a grid-point array in which not all variables are held at the same locations. For example, the variables p, T, ρ, q, and w may be held at points marked with crosses whilst u and v are held at the staggered points marked with dots (figure 14.4). Such staggered grids are designed to make the application of finite-difference approximations either simpler or more accurate or both. On the staggered grid shown above the pressure-gradient force must be calculated at the points marked with dots where u and v are held. For example at $(i + \frac{1}{2}, j + \frac{1}{2})$

$$\left(\frac{\partial p}{\partial x}\right)_{i+1/2,\,j+1/2} = \frac{1}{a}\left[\frac{p_{i+1,\,j} + p_{i+1,\,j+1}}{2} - \frac{p_{i,\,j} + p_{i,\,j+1}}{2}\right]. \qquad (14.5)$$

Notice that in this case the finite-difference approximation is calculated over a single grid length a; also that averaging has been employed to obtain values of p at the required positions. Such averaging is a common feature of finite difference equations in numerical models.

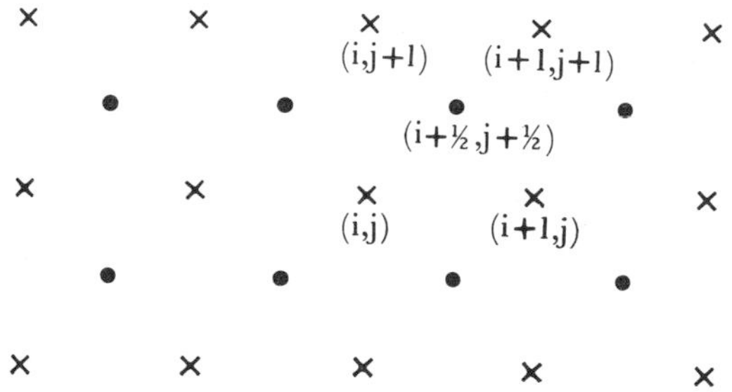

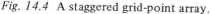

Fig. 14.4 A staggered grid-point array.

In the diagnostic equations, the replacement of spatial derivatives with finite differences enables grid-point values of the diagnosed variables to be calculated from grid-point values of the predicted variables. In the predictive equations, the finite differences lead to expressions which may be used at each grid point to calculate the time rates of change of each of the predicted variables. The calculation of these rates of change represents the first stage in the numerical solution of our equations.

Integration in time

Once the rate of change of each predicted variable is known at each grid point through the application of finite-difference approximations, the next requirement is for a method of integration in time so that the state of the model atmosphere may be advanced. The first method that will come to mind is probably the forward integration scheme. In this scheme, the rates of change are multiplied by the chosen interval of time (called the time step) and the resulting increments are added to the original grid-point values to obtain new values after the elapse of one time step. Thus, if q_0 is the original value of q at some grid point and $(\partial q/\partial t)_0$ its calculated rate of change, then the value q_1 after one time step (δt) would be

$$q_1 = q_0 + \delta t \left(\frac{\partial q}{\partial t}\right)_0. \tag{14.6}$$

After the entire state of the model atmosphere at the end of the first time step has been calculated, the process may be repeated to obtain q_2. After n time steps

$$q_n = q_{n-1} + \delta t \left(\frac{\partial q}{\partial t}\right)_{n-1}. \tag{14.7}$$

This simple picture of step-by-step integration is complicated by a phenomenon known as computational instability. It may be shown through mathematical analysis that the combination of a particular time integration scheme with a particular finite-difference approximation applied to a particular differential equation may lead to numerical solutions in which the numbers grow rapidly in magnitude, departing disastrously from the true solution, and soon become too big for the computer to handle. We usually say that the integration has 'blown up'. It may be shown that the forward integration scheme with centred finite differences in space applied to an equation representing advective changes (Panofsky, 1981b) is *always* computationally unstable.

Luckily, an alternative integration scheme is readily available in which our calculations are centred in time. In this case,

$$q_n = q_{n-2} + 2\delta t \left(\frac{\partial q}{\partial t}\right)_{n-1}, \tag{14.8}$$

and for obvious reasons this is known as the leapfrog scheme. This scheme together with centred spatial differences is stable for an advection equation *provided that* the dimensionless parameter,

$$\frac{U\delta t}{a},$$

(where U is the magnitude of the advecting velocity) is less than some critical value. We see that for a given grid length, a, and a given maximum advecting velocity this stability criterion sets some limit on the time step δt which may be used.

By repeated application of the leapfrog method the state of the model atmosphere may be advanced through any desired forecast period. Other precautions may be needed to prevent the integration being spoiled by more subtle and slowly acting forms of numerical instability.

Besides the leapfrog scheme a number of other integration schemes are available and are used according to taste and application. If, like the leapfrog scheme, they are *explicit* methods (in which values for time step n may be locally calculated at each grid point from values at previous time steps) they will be subject to similar restrictions on δt to that described above. If, however, they are *implicit* methods (in which the state of the model atmosphere at time step n must be obtained by solving a set of simultaneous equations for all the grid points together) they may be free from such restrictions, although this advantage will be partly offset by increased computation per time step.

Given the large number of grid points in most models, the seven variables, and the necessity of short time steps to maintain computational stability, it is not difficult to see why numerical modelling imposes a heavy computational burden.

Thus, the computational efficiency of an integration scheme is an important factor in view of the limited computer resources available. Recently, considerable effort has been devoted to the development of more efficient integration schemes for use in numerical weather prediction.

Physical processes

Our basic equations include terms which represent the effects of physical processes such as radiative exchange, phase changes of water substance, and exchanges between the atmosphere and the underlying surface. These terms describe sources or sinks of momentum, energy, or water substance. The processes represented are often very complex, but for the purposes of numerical modelling must be sufficiently simplified or parameterized so that their effects may be calculated from the available grid-point variables.

The effects of a physical process are sometimes calculated as contributions to the rates of change of one or other of the seven meteorological variables, and thus may be included in the chosen integration scheme. Usually, however, they have little impact on the numerical stability of the integration scheme, and often, to improve computational economy, they are recalculated at intervals of, say, an hour rather than every time step.

Alternatively, the physical process might be represented by an adjustment procedure. Take as an example the condensation of water vapour. In most numerical models the changes in temperature and humidity are calculated each time step without reference to the possibility of condensation. Then, after each time step and every grid point, the new value of the humidity mixing ratio (q) is compared with the saturation mixing ratio $q_S(T, p)$ which corresponds to the calculated values of temperature (T) and pressure (p). If $q \leqslant q_S$ then nothing further need be done. But if $q > q_S$ the usual assumption is that condensation of water vapour was taking place at that grid point during the preceding time step at a sufficient rate to keep $q = q_S$. In this case, a further calculation is carried out to determine adjusted values of q and T which leave the air exactly saturated and take correct account of the latent heat released. The surplus water vapour which has been condensed becomes the model's cloud water or precipitation.

Subgrid-scale processes

A fundamental and inevitable problem associated with grid-point modelling concerns the role of motions in the real atmosphere which operate on a scale smaller than that resolved by the finite-difference (or spectral or finite element) mesh. Some such subgrid-scale motions can be happily ignored, but others have an important impact on the larger-scale motions explicitly represented in the model. A notable example here is cumulus convection as the subgrid-scale process in numerical weather prediction models. Some kind of representation of the vertical heat redistribution achieved by cumulus convection is absolutely necessary in such models. As with physical processes, the representation may be in the form of parameterized rates of change or, more commonly, through an adjustment procedure which insists that some critical lapse rate of temperature is not exceeded.

Although of differing basic natures, subgrid-scale processes like cumulus convection and the physical processes outlined earlier tend to be treated together in many numerical models. In an imprecise but useful and much used nomenclature, they are together referred to as a model's 'physics'. This is in contrast to so-called 'dynamics', which indicates the application of finite differences and the integration scheme to the equations of motion, thermodynamics and continuity.

Subgrid-scale motions similar in principle to cumulus motions, but usually handled differently in models, are those represented by eddy diffusion terms in the predictive equations. These terms have a dual role; in part they represent the poorly understood effect of the subgrid-scale eddies; but also they bring about a desirable (or even necessary) degree of smoothing of the grid-point variables, thereby perhaps avoiding some form of numerical instability or at least making the appearance of computed fields more acceptable. In some models, a numerical filtering operation is carried out at intervals to remove 'noise' from the integrations; this may be either additional to or in place of the diffusion terms in the basic equations.

Initial and boundary conditions

The kind of time integration of the basic equations which has been described in this chapter is based on a mathematical approach to the study of the atmosphere. In mathematical parlance, we have approached the motion of the atmosphere as an initial-boundary-value problem. To say this is merely to make two obvious statements. First, that in order to carry out a step-by-step integration we must specify the initial conditions (i.e. the grid-point value of every variable at the starting time). Secondly, that in order to use finite-difference approximations we must apply some suitable boundary conditions at the edges of our grid array (where the finite differences cannot be evaluated). Some numerical models use idealized, mathematically specified, initial data, but more often a consideration of initial conditions leads us into the realm of the analysis of observational data. In particular we are led to 'objective analysis', that is to techniques for the automated calculation of grid-point values from observations (Pedder, 1981c). This is a big subject in its own right. Another important matter concerns the balance between wind and pressure fields in the initial data. If, due to observational or analysis shortcomings, the initial fields are not balanced (roughly speaking, if the geostrophic approximation is seriously violated) the subsequent integration will be ruined. It was the neglect of this initial balance that ruined L. F. Richardson's famous first attempt at numerical modelling. The computations carried out to ensure an initial balance are often referred to as 'initialization'.

Boundary conditions at the top of our model atmosphere and at the earth's surface form a very important component of the numerical modelling concept. In sigma co-ordinates these upper and lower boundary conditions are simply that the vertical velocity is zero, but in other co-ordinate systems matters may be more complicated.

Attention must also be given to lateral boundary conditions. If our numerical model deals with the entire globe there are, of course, no lateral boundaries. But, in other cases, boundary values must be held constant, or changed according to some theoretical framework, or specified from some external source.

Historical note

Mention has already been made of the pioneering attempt at numerical modelling by Richardson around the time of the First World War. Richardson's work eloquently makes the point that numerical modelling has an existence independent of electronic computers. But without computers the task is almost impossibly arduous, so that rather little further development took place until suitable computers began to become available shortly after the Second World War. Richardson had worked with the basic equations in their 'primitive' form (much as we have considered them here) but the later work concentrated on a derived set of equations, known as the 'filtered' equations, in which vorticity was the main predicted variable. This approach avoided the problem of balanced data which had defeated Richardson, and also allowed great savings in computer time. In filtered models, however, it is difficult to represent physical processes realistically, and so eventually, from about 1960 onwards, attention turned back to the primitive equations. The problem of balanced data was resolved in various ways, whilst recent advances in the design of integration schemes have allowed primitive equation models to become at least as economical in computing time as filtered models.

Coincident with the return to the primitive equations came a widening of the applications of numerical modelling in meteorology. The filtered models were essentially numerical weather prediction models; they usually dealt with a limited portion of the earth's surface, produced forecasts for a day or two ahead, and had sufficient resolution for cyclones and anticyclones to be represented. These were the forerunners of today's primitive equation numerical weather prediction models. But also today we have, on the one hand, global models of the general circulation of the atmosphere with their potential applications to long range forecasting, climate and climatic change studies. On the other hand, numerical modelling has been extended into studies of smaller-scale phenomena; fronts, sea breezes, cumulonimbus and so on.

The development of numerical weather prediction models is generally acknowledged to have been of significant benefit for weather forecasts one, two and three days ahead. It remains to be seen whether in the years ahead comparable impact can be made on problems involving both longer and shorter-time scales. There are certainly severe problems to be faced, but since what could only be a dream for L. F. Richardson in 1922 has now largely been accomplished, we may reasonably expect further exciting developments in the future.

References

Atkinson, B. W. (1981a) 'Atmospheric waves', this volume, 100–115.
Atkinson, B. W. (1981b) 'Dynamical meteorology: some milestones', this volume, 116–129.
Harwood, R. S. (1981) 'Atmospheric vorticity and divergence', this volume, 33–54.
Panofsky, H. A. (1981a) 'Atmospheric hydrodynamics', this volume, 8–20.
Panofsky, H. A. (1981b) 'Atmospheric thermodynamics', this volume, 21–32.
Pedder, M. A. (1981a) 'Practical analysis of dynamical and kinematic structure: principles, practice and errors', this volume, 55–68.
Pedder, M. A. (1981b) 'Practical analysis of dynamical and kinematic structure: some applications and a case study', this volume, 69–86.
Pedder, M. A. (1981c) 'Practical analysis of dynamical and kinematic structure: more advanced analysis schemes', this volume, 87–99.

Further reading

Haltiner, G. J. (1971) *Numerical Weather Prediction*, New York, Wiley.
Richardson, L. F. (1922) *Weather Prediction by Numerical Process* (republished 1965), New York, Dover.
Thompson, P. D. (1961) *Numerical Weather Analysis and Prediction*, London, Collier-Macmillan.

15

Epilogue: a perspective of dynamical meteorology

J. SMAGORINSKY
Princeton University

We have seen from the preceding chapters how the basic physical laws which govern the workings of the atmosphere have been deduced. These laws, the equations of motion, the first law of thermodynamics, and the mass continuity equation, generally take the mathematical form of partial differential equations (PDE's). The exception is the semi-empirical 'perfect' gas law, i.e., the equation of state, which is algebraic. We have also seen specific examples of how some mathematical techniques, such as spectral analysis and perturbation methods, have been developed to help translate these laws into a more readily useful form, that is, to find solutions to the PDE's in special cases.

Let us try to collect our thoughts to form an overall perspective.

The equations of motion are valid for describing the momentum laws governing all phenomena in a continuous medium (see, for example, Einstein, 1955) which includes all liquids and all gases of sufficiently high density that molecular interactions can be expressed in the form of pressure forces. Apart from the fact that this demonstrates a universality and commonality of the underlying physics of all fluids on this planet and elsewhere, a statement of the equations of motion does little to tell us why different phenomena might be expected to appear in a teacup, in a stream, or in a planetary atmosphere.

Planetary atmospheres

In meteorology, we are interested in the motions of the atmosphere relative to the rotating spherical planet. The mathematical formalism for transforming the equations of motion of a fluid from an absolute framework is straightforward, and this gives rise to the appearance of Coriolis forces in the transformed equations. However, again, as those PDE's stand, they do not enlighten as to why in our atmosphere there should exist cumulus clouds, tornadoes, hurricanes, extra-tropical cyclones (or depressions) or jet streams to the exclusion of other conceivable, though not observable, phenomena. Between the size of a cloud droplet and the circumference of the Earth there is a possible range of sizes of 11 orders of magnitude, that is from 1 mm to 10^5 km (figure 15.1). Why is it that hurricanes of horizontal dimensions much less than 10^3 km are not to be found and why are they more common in the north Pacific Ocean than in the north Atlantic? What determines the narrow range of observed sizes of cumulus clouds? Why do extra-tropical cyclones not last for several months rather than several days or a week? Why do they tend to be ubiquitous in the winter hemisphere, whereas tornadoes are rare in all seasons?

The reasons why certain phenomena occur and why they have narrowly confined characteristic dimensions in time and space are the task of dynamical meteorology to divulge. For this we must get beyond the general PDE's and into some tractable form which can isolate the classes of specific phenomena we wish to understand. This is done in stages. It will be useful to refer to a schematic view of the relationship of these stages in discussing the methodology (figure 15.2). We have already alluded to the first stage: the transformation of the general conservation laws to a form which expresses the physical constraints relative to a rotating sphere will assist us in isolating phenomena to be found in a planetary atmosphere. What distinguishes the atmosphere of Jupiter from that of Earth in the equations? To begin with, there are a number of constants or parameters that appear in these equations (figure 15.3). First, there are those which depend upon the properties of the planet itself: its radius a, its mass or gravitational acceleration g, and its rotation rate Ω. The latter determines the strength of the Coriolis force and the length of the day. Then, there are fundamental parameters of the atmosphere itself: its mass or the average pressure p_* it exerts at the planet's surface, and the specific heats of the gaseous mixture both at constant pressure c_p and constant volume c_v (which together determine the 'gas constant'). It is the most massive gaseous constituents that determine c_p and c_v. For the Earth, these gases are molecular nitrogen, N_2, and oxygen, O_2. The radiative properties of the atmosphere are determined by the absorptive characteristics of its constituents with respect to the sun's spectrum of radiation $S_\infty(\lambda)$, not only in its visible part but at its all-important fringes in the infra-red (long-wave radiation) and the ultra-violet (short-wave radiation). On Earth, the most radiatively active constituents are the trace gases – carbon dioxide, CO_2, ozone, O_3, and water vapor, H_2O. The other constituents of the gaseous atmosphere, argon, neon, krypton, xenon, hydrogen, and radon, have no consequence in their massiveness or on the radiative transfer. In contrast, on Venus and Mars the most massive constituent, CO_2, is also the most radiatively active.

Other, non-gaseous atmospheric constituents, such as clouds and dust, can also profoundly influence how the planetary atmosphere will react to the radiation from the sun and to the important secondary infra-red radiation going outward from the planet's surface.

Then there are the orbital characteristics of the planet. The inclination of the planetary axis to the plane of the ecliptic determines the nature of the seasonal variations and the orbit itself determines the distance from the sun and the length of the year. Does the planet itself give off appreciable heat, say compared to the incident solar insolation? Saturn and Jupiter presumably do.

Finally, the characteristics of the planet's surface can be important: how

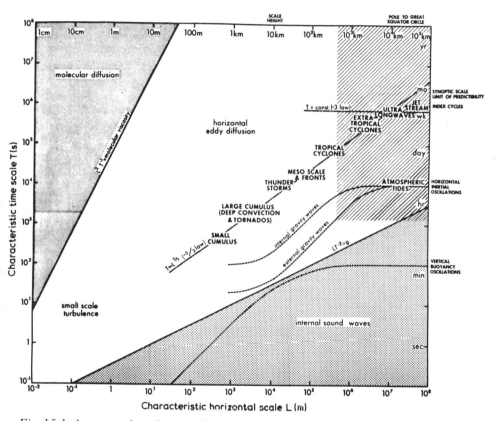

Fig. 15.1 A space–time domain for characteristic atmospheric phenomena. The unstippled region encompasses most of the kinetic energy containing phenomena, with the predominance of extra-tropical cyclones, ultralong waves, and the jet stream. The cross-hatched area denotes scales and phenomena typically resolved by general circulation models (based on Smagorinsky, 1974).

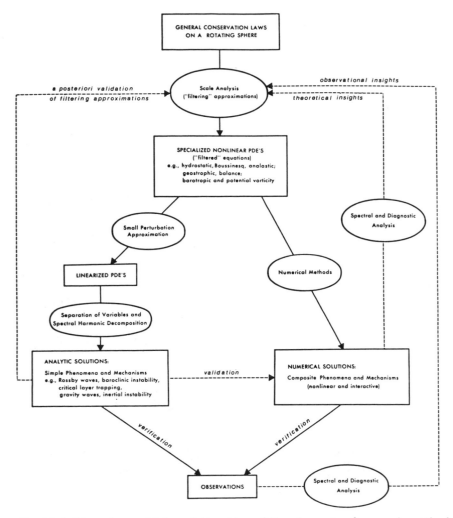

Fig. 15.2 The roles and inter-relationships of the elements of research methodology in the development of dynamical meteorology.

rough is it, Z_* (this may vary between field, forest and sea); how does it reflect
sunlight, A_* (this will vary between desert, snow surface and sea); how easily
does it store heat, K (the desert and the sea are extremes); can the surface store
water that might be available for evaporation?

Not all of these characteristics are equally important in influencing the
characteristics of each of the phenomena that are possible in a planetary atmosphere. For example, intuitively we might suspect that small-scale atmospheric
turbulent elements of several centimetres in size and with lifetimes of the order

ORBITAL DATA

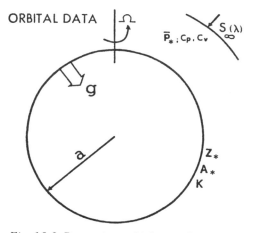

Fig. 15.3 Parameters which must be specified numerically in the physical conservation laws governing a planetary atmosphere.

of seconds might not be directly affected by the fact that the Earth is rotating. But, in most other cases, it is not immediately obvious and one of the key facets of developing an understanding is to determine when certain parameters are important, why and how.

We are still left with a very complex system of laws even after all of these specializing parameters have been prescribed as actual numbers. The most general planetary equations are still valid for a tremendously varied range of scales and phenomena such as exist in figure 15.1.

Scale analysis

What suggests itself is that somehow the equations could be simplified by searching for clues as to when certain parameters are important for certain classes of phenomena. The clues often come from direct observation of the atmosphere itself. An example has already been given earlier in connection with the likely insensitivity of small turbulent vortices to the fact that the Earth is rotating.

Another example is the following. The characteristic depth of the atmosphere can be estimated by noting that 90 per cent of its mass lies below about 15 km and 99 per cent lies below about 30 km. This characteristic vertical scale of a few tens of kilometres (the 'scale height' of the terrestrial atmosphere) is small compared to the horizontal dimensions of an extra-tropical depression, an observational fact. We would then suspect that the vertical velocity of the air may be much smaller than the horizontal velocities and that the vertical accelerations may also be small compared to the gravitational acceleration. Applying this assertion to the vertical equation of motion reduces it to a diagnostic relationship, the 'hydrostatic equation'. We would expect the hydrostatic equation to be equally valid to the study of extra-tropical cyclones as is the general time-dependent vertical equation of motion. However, the advantage is that it is unnecessary to carry the added complexity necessary to insure that the solutions be equally applicable to atmospheric phenomena, where the horizontal scale is comparable to the vertical scale, for example cumulus clouds or vertically propagating sound waves.

It is therefore useful somehow to discover in advance the specialized versions of the general physical PDE's valid for different parts of figure 15.1. As was said earlier, an essential part of weaving a fabric of understanding consists of isolating special cases. When an *a priori* simplification is made in the system of PDE's, it must be considered tentative until its validity can be verified *a posteriori* in the solutions themselves. The ultimate measure is the observed physical medium itself, when possible.

The systematic search and rationale for an *a priori* classification of the specialized PDE's is called a 'scale analysis'.

As has already been pointed out, some intuitive aspects of scale analysis are immediately applicable to the specialization and the simplification of the governing physical equations toward relevance to a subset of physical phenomena. But often subtle ramifications cannot easily be inferred. To illustrate this point, we note that the hydrostatic approximation carries with it secondary consequences, that have to be taken into account for consistency (Smagorinsky, 1958). The hydrostatic approximation in itself renders incongruous the energy properties of the equations of motion. To restore consistency, namely to insure that the change of the kinetic energy of a parcel of air depends only on the work done by the pressure gradient and external forces, one must then also modify the Coriolis terms in the horizontal components of the equations of motion and, furthermore, must require that the height of an air parcel above the Earth's surface is always small compared to the radius of the Earth. The complete set of physical conservation laws for a quasi-hydrostatic system in which water vapour is thermodynamically active, is shown schematically in figure 15.4. It is also known in this form as the 'primitive equations.'

A classical attempt to systemize scale analysis for the study of the large-scale motions of the atmosphere was proposed by Charney (1948). He did this by assigning observationally typical large-scale values to the dependent variables (horizontal wind and temperature and their vertical gradients) corresponding to the independent variables which characterize the time—space scale span of interest. The resulting simplifying or 'filtering' approximations eliminated possible solutions which would correspond to meteorological 'noise'. In this way he was able to derive consistently, though not for the first time, the 'vorticity equation'*, the 'geostrophic approximation' and the property that the horizontal

* The notion that a fluid can have a property which we call vorticity is not particularly enlightening in itself. However, the fact that this property has special significance, for example that under certain circumstances it is a conservative property, in the same way as mass and energy, can only be deduced from the set of governing laws, in this case the vorticity equation.

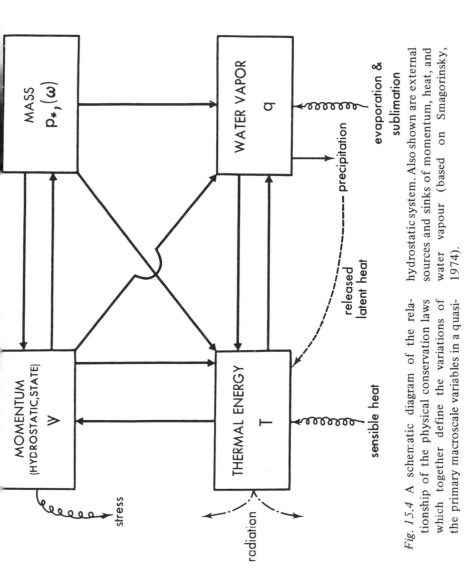

Fig. 15.4 A schematic diagram of the relationship of the physical conservation laws which together define the variations of the primary macroscale variables in a quasi-hydrostatic system. Also shown are external sources and sinks of momentum, heat, and water vapour (based on Smagorinsky, 1974).

Dynamical Meteorology

divergence is small but significant. In fact, as a consequence, the inequalities

$$f > |\zeta| > |D| > \frac{1}{p_*}\int_0^{p_*} Ddp$$

can be shown to hold. The Coriolis force, f, in middle latitudes is about 10^{-4} s^{-1}; the vertical component of relative vorticity ζ is of order 2×10^{-5} s^{-1}; the horizontal component of divergence D is about 5×10^{-6} s^{-1}; and the vertical integral of D is even smaller, about 10^{-6} s^{-1}. The reason for the latter is that the large-scale dynamics tend to compensate low tropospheric divergence with high tropospheric convergence and *vice versa*. This was already deduced in 1912 by Dines (Atkinson, 1981b). The net result gives rise to relatively small pressure changes at the Earth's surface. We now know that this reflects the fact that the energy of the long external gravity waves in the terrestrial atmosphere is much less than that of the long quasi-geostrophic 'Rossby waves', the latter having the property of tending to conserve absolute vorticity. But Margules at the turn of this century already realized that an attempt to predict the surface pressure changes by using observed winds in the 'pressure tendency' equation, which is derivable from the hydrostatic and continuity equations, is doomed to failure (see for example Exner, 1925). Another way of putting it is that if one decomposes the observed large-scale horizontal wind into a rotational and a divergent part, the latter has far less energy than the former in the extra-tropics so that the accuracy of even modern-day wind measurements is inadequate to detect adequately the divergent part of the wind in mid-latitudes. These inequalities therefore are important in what they imply about the relative importance of different atmospheric mechanisms.

Over the past quarter of a century, means for further rationalizing scale analysis have been sought, with the object of replacing some of the *ad hoc* intuitive methodology by a more universal approach. For example, is it possible to derive the hydrostatic approximation and its secondary consequences at the same time?

The basic philosophy derives from the method of dimensional analysis that has proven so useful in physical science in the construction of valid scaled-down laboratory analogues of natural physical systems (see, for example, Birkhoff, 1950). The classical examples come from the studies of turbulence in the wind tunnel and of the aerodynamic properties of model airfoils. The key was the discovery that when the governing physical laws are written in non-dimensional form, certain ratios of forces appear as coefficients. In the case of laboratory analogues, these coefficients, such as the Reynolds number (the ratio of inertial to viscous force) and the Froude number (the ratio of inertial to gravitational force), should be the same in the model as it is in its prototype to insure dynamical similitude, that is for the analogue to be valid. In the case of scale analysis, the sizes of the repertoire of these non-dimensional numbers determines which forces and mechanisms are primary in different parts of the time—space domain. Since they are the coefficients in the PDE's, the conditions under which they are small determine which of the terms can be neglected, thus simplifying the operative PDE's–the 'filtered equations'.

In addition to the Reynolds, *Re*, and Froude, *F*, numbers one encounters the Richardson number, *Ri* (the ratio of buoyancy to vertical shearing force), the Rossby or Kibel number, *Ro* (the ratio of inertial to Coriolis force) and the Ekman number, *E* (the ratio of viscous to Coriolis force). So, for example, the large-scale, quasi-horizontal motions are characterized by $Ro \ll 1$ and $Ri \gg 1$, while in a tornado $Ro \gg 1$. In the planetary boundary layer $E \sim 1$, but in the free atmosphere $E \ll 1$, and within a few metres of the Earth's surface sublayer, $Ro \ll E \gg 1$.

In this way, one can derive the specialized set of laws valid for a restricted part of the space–time spectrum or for a special part of the atmosphere, such as in the boundary layer or for the equatorial tropics.

The main complexity at this point is mathematical and it reflects an important physical fact. The partial differential equations are 'non-linear' and, in general, no analysis can be written down which can express how a particular dynamical property or phenomena intrinsically describable by the PDE's depends on the parameters and on the boundary and initial conditions. The reason is that, as the equations stand, all the phenomena can co-exist and, in principle, interact with each other. On the face of it, this is a desirable property of the non-linear equations but, in practice, it renders the extraction of a simple understanding difficult, if not impossible, in this form. Means are needed to unravel further the inherent generality and complexity to a more manageable form.

Before going on, it should be said that the equations in this form, although not generally solvable by analytical methods, can be solved approximately by 'numerical' methods through the application of large computers. This is a field which has been much exploited in the past 25 years. However, despite the tremendous advantages of numerical methods, in themselves they are no clear substitute for analytical solutions, if such solutions are possible at all. Both analytical methods and numerical methods each have their individual advantages and disadvantages. The two together have in recent years proven to be a formidable and unique combination (see, for example, Miyakoda, 1974; Smagorinsky, 1974). However, even before computers, progress was possible and in fact analytical dynamical meteorology laid the foundations for an intelligent and orderly approach to the exploitation of numerical methods (see Smagorinsky, 1972).

Linearization

A key for mathematical simplification lies in the method of 'small perturbations' which makes use of the fact that we may be able to divide a complex atmospheric condition into two parts such that they do not fully interact (Atkinson, 1981a). If one of the parts, the perturbation, is relatively smaller in some sense, say in its energy, than the basic state, one can assume that the main influence is from the larger to the smaller. For an analogy, consider the orbits of the Earth and Moon. The Moon, because its mass is considerably less than that of the Earth, hardly influences the Earth's orbit. In fact, the moon's orbit is almost completely determined by that of the Earth. Hence, we see that such a separation, when valid, permits one to reduce a formally highly interactive problem to one which, in the first approximation, is causal, that is we can usefully try to distinguish which came first, the chicken or the egg. A very useful step!

A secondary advantage is that the effects of a number of such different perturbations can be added together to give an approximation of their composite characteristics. That is, they satisfy the condition of 'superimposability'. In terms of our analogy, this could be the case of more than one small moon. The analogy reveals that some care must be used in superimposing effects. For example, when the moons are sufficiently close to each other, their mutual gravitational attraction is greater than is the Earth on either.

The small perturbation approximation then permits a linearization of the PDE's and suddenly makes available a tremendous body of mathematical analysis from which the solutions can be expressed in terms of known mathematical functions, the 'analytical' solutions. Outstanding among these is the method of 'spectral decomposition', which states that within certain constraints of mathematical continuity, any arbitrary function of space and time, such as the pressure or temperature field, can be represented by a spectral harmonic series. The most common example is a trigonometric series of waves of different wavelengths and periods. By the small perturbation approximation, each term may be considered separately and the corresponding linear solution for each sought. By the superimposability property, the sum of an arbitrary number of such solutions, corresponding to different wavelengths and periods, is also a solution. This permits one to consider fairly complex initial and boundary conditions which can be synthesized from a spectral harmonic series.

Instability

A slight generalization provides a greatly enlarged analytical tool. Let us assume that the spectral frequency, which is inversely proportional to the period, can be expressed as a complex number. That is, the frequency is not restricted to being just a real number, but may also have an imaginary part. The solutions of the linearized equation will in general yield relationships between the frequency and the wavelength, the 'frequency equation'. If these solutions say that the imaginary part of the frequency must be zero, then an initially wave-like disturbance must remain wavelike, that is it is 'neutrally stable'. If, on the other hand, the solution is found to be complex, then it generally tells us that under certain circumstances the original wave-like solution may grow or decay exponentially, respectively 'instability' or 'stability'. In the case of exponential growth, the small perturbation approximation ultimately becomes invalid, that is 'non-linear'. But it does tell us under what circumstances a physical instability is possible. The significance of this step cannot be overemphasized. Small non-measurable perturbations of all wavelengths and periods are being excited everywhere in the atmosphere all of the time, and yet figure 15.1 shows us that we see only a small selection of phenomena almost discretely separated from each other in their characteristic horizontal size and characteristic lifetime. In fact, those that we do see are mainly the result of instabilities. For example, linear perturbation theory tells us that an instability, called 'baroclinic instability', can occur under certain circumstances that will yield large disturbances (extra-tropical depressions) with characteristic horizontal dimensions of several thousand kilometres and with an amplitude doubling time of several days. This is the primary mechanism for potential to kinetic energy transformation in extra-tropical latitudes. Smaller and larger perturbations will not grow as rapidly or at all. Physically, the growth rate will depend primarily upon the buoyant stability of the atmosphere, the vertical wind shear and the Earth's spherical geometry, that is the value of the Coriolis force and its rate of variation with latitude. Baroclinic instability was independently discovered theoretically by J. G. Charney (1947) and E. T. Eady (1949) in the later 1940s.

Similarly, stability analysis can be applied to a great variety of problems, for example the 'barotropic instability' of an atmospheric current with horizontal shear. On the other hand, some types of instability cannot form from small perturbations; they must be forced into a large or finite amplitude before the destabilizing process can cause further growth. Examples are the instability of internal gravity waves on the generation of deep oceanic turbulence, or the buoyant instability of a 'conditionally unstable' moist atmosphere to form cumulus clouds. In the latter case one way that the finite amplitude can be established is by forcing the moist air up the windward side of a mountain thereby cooling, because of adiabatic expansion, until eventually the air becomes saturated. Another is when the air near the Earth's surface is forced to rise by the heating of the ground as the result of daytime solar insolation.

One can find a simple mechanical analogy by considering a cone (figure 15.5). Placed on its side, it is neutral to both small and large rolling displacements or perturbations. Placed on its apex, its centre of gravity is unstable to any small perturbation. Placed flat on its base, it is stable to small perturbations. But if tipped over far enough by an external force, a finite amplitude perturbation, it will continue to flip over. This then is a conditional instability. Finite amplitude instability cannot be directly deduced from small perturbation analysis.

Some further comments on the evolution of ideas

It seems to be a fact of science that many notions are rediscovered a number of times. Sometimes, an idea is so novel that, upon its first exposure, it goes unaccepted and must await a more hospitable state of the science when new, more definitive, observations have been acquired, or when a better-developed underlying body of theory has been developed. By then, the original idea may have been forgotten and therefore goes unacknowledged. In other cases, two workers will independently and by different means obtain essentially the same result, well illustrated by the discovery of baroclinic instability by Charney and Eady. Simultaneous discovery usually occurs when the antecendent developments are ripe to be exploited in what seems to be the next logical step, which may of course by a very large step.

Then again, a correct idea may be voiced early, but perhaps for the wrong

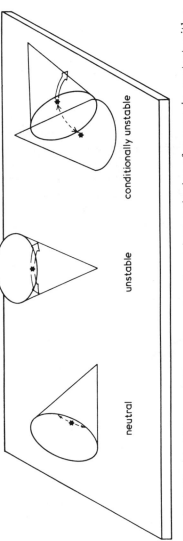

neutral   unstable   conditionally unstable

Fig. 15.5 The stability conditions appropriate to various attitudes of a cone in contact with a plane. The star indicates the centre of gravity of the cone.

reason, and therefore lies dormant to be rediscovered. An interesting illustration of the latter situation occurs in an 1872 paper by a French mining engineer, Peslin, which was translated by Cleveland Abbe in the first of his three marvellous volumes of collected works. In discussing the role of the Earth's vorticity, Peslin concludes that the Coriolis force due to vertical motion is probably negligible compared to that due to horizontal motions, except possibly 'in storms and hurricanes'. This is a consequence that can be deduced as energetically consistent with the hydrostatic approximation, as we discussed earlier. But Abbe points out in a footnote that 'It will be perceived that, in these formulae, Peslin treats of the rotation of the earth [sic] to the neglect of the rotation of storm-winds about their centres'. An apparent allusion, perhaps accidental, that $f > |\zeta|$. Incidentally, Peslin also discerned the relevance of his penetrating visions to ocean currents, in a sense a precursor of the unity of geophysical fluid dynamics.

One can find many other remarkable insights in the literature of the latter nineteenth and early twentieth centuries, gleaned from pitifully sparse observational evidence and a very primitive theoretical base.

In the introduction to this third collection, Abbe (1910) remarks, 'The modern sounding balloon has assured us of the intimate connection between the lowest stratum of air and that which is 20 miles above us; but the conditions above this latter level are doubtless of equally great importance to our surface climatology . . .'. This was a realization that was to become more cogent with the advent of copious radiosonde measurements in the 1940s. But there are still some today that feel that the climate at the Earth's surface can be understood without regard to the structure and dynamics of the atmosphere's general circulation.

In the same introduction, Abbe suggests that 'It only remains for future students to combine the equations of thermodynamics with those of hydrodynamics so as to further elucidate the details of the phenomena as to time and place – a result that we may hope will eventually be attained by the analysis of fields of force that is now being perfected by Bjerknes of Christiania [Oslo].' Indeed, V. Bjerknes (1904) and then L. F. Richardson (1922) during the first World War laid a blueprint for carrying this idea forward. But it was not until after World War II that the plan came to fruition as 'numerical weather prediction'. It required an evolution of the theoretical base, that is an understanding of the existence and dynamics of Rossby waves as solutions of the quasi-geostrophic vorticity equation; it required a good net of aerological observations to establish initial conditions in mid-troposphere; it required new methods for numerically integrating the vorticity equation (Richardson's original method for the primitive equations was actually in error); it required the modern electronic computer.

One cannot resist calling attention to a few more early intelligent grasps. One of the giants of the important later nineteenth century, German–Austrian school of meteorology was Wilhelm von Bezold, mainly known for his fundamental contributions to atmospheric thermodynamics. Neuhoff (1900) pointed out that 'von Bezold also first called attention to the fact that the processes going on in the atmosphere are often not reversible except in a very limited sense'.

But even more subtle a spontaneous understanding came from F. H. Bigelow

(1903). He asserted 'The cyclone is not formed from the energy of the latent heat of condensation, however much this may strengthen its intensity; it is not an eddy in the eastward drift, but is caused by the counterflow and overflow of currents of air of different temperatures'. Bigelow apparently sensed that the instability process responsible for the inception of mid-latitude depressions, which we now know as baroclinic instability, is possible in a dry atmosphere. The availability of latent energy in the form of moisture would serve only as a modifying agent. Of course, we now know considerably more about the nature of that modification. For example, moist dynamics give rise to smaller-scale and, surprisingly, less intense depressions. The latter apparent paradox is due to the fact that baroclinic instability is actuated by the global poleward heat transport requirements of the meridional radiative gradient. If the baroclinic eddies can, at the same time, transport heat in latent as well as sensible form, the eddies themselves can be less intense.

Acknowledgement

I wish to acknowledge my gratitude to Professor John M. Wallace for his helpful suggestions.

References

Abbe, C. (1910) 'The mechanics of the earth's atmosphere', 3rd Collection, *Smithson. Misc. Coll*, 51 (4).

Atkinson, B. W. (1981a) 'Atmospheric waves', this volume, 100–115.

Atkinson, B. W. (1981b) 'Dynamical meteorology: some milestones', this volume, 116–129.

Bigelow, F. H. (1903) 'Studies on the meteorological effects of the solar and terrestrial physical processes', *Weather Bureau Pub.*, 290 (1950), Washington, DC, p. 37.

Birkhoff, G. (1950) *Hydrodynamics, A Study in Logic, Fact, and Similitude*, Princeton, Princeton University Press.

Bjerknes, V. (1904) 'Das Problem von der Wettervorhersage, betrachtet vom Standpunkt der Mechanik und der Physik', *Meteor. Z.*, 21, 1–7.

Charney, J. G. (1947) 'The dynamics of long waves in a baroclinic westerly current', *J. Met.*, 4, 135–62.

Charney, J. G. (1948) 'On the scale of atmospheric motions', *Geofys. Publikasjoner*, 17 (2).

Dines, W. H. (1912) 'The free atmosphere in the region of the British Isles', *Geophysical Memoirs*, No. 2, London, Meteorological Office.

Eady, E. T. (1949) 'Long waves and cyclone waves', *Tellus*, 1, 35–52.

Einstein, A. (1955) *The Meaning of Relativity*, 5th edn, Princeton, Princeton University Press, p. 19.

Exner, F. M. (1925) *Dynamische Meteorologie*, 2nd edn., Vienna, Springer-Verlag, p. 75.

Miyakoda, Kikuro (1974) 'Numerical weather prediction', *Am. Sci.*, 62(5), 564–74.

Neuhoff, O. (1900) 'Adiabatic changes of condition of moist air and their determination by numerical and graphical methods', *Mem. R. Pruss. Met. Inst.*, 1 (6), 271–306; translated by C. Abbe (1910), 'The mechanics of the earth's atmosphere', 3rd Collection, *Smithson. Misc. Coll.*, 51 (4), 430–85.

Peslin, M. (1872) 'On the relation between barometric variations and the general atmospheric currents', *Bull. int. de l'Obs. de Paris et l'Obs. Phys. Cent. de Montsouris*, May 26 to July 7, inclusive; translated by C. Abbe (1878) *Short Memoirs on Meteorological Subjects*, Smithsonian Report for 1877, Washington, D.C., Government Printing Office, 465–78.

Richardson, L. F. (1922) *Weather Prediction by Numerical Process*, Cambridge, Cambridge University Press.

Smagorinsky, J. (1958) 'On the numerical integration of the primitive equations of motion for baroclinic flow in a closed region', *Monthly Weather Rev.*, 86 457–66.

Smagorinsky, J. (1972) 'The general circulation of the atmosphere', *Meteorological Challenges: A History*, Ed. D. P. McIntyre, pp. 3–41, Ottawa, Information Canada.

Smagorinsky, J. (1974) 'Global atmospheric modeling and the numerical simulation of climate', *Weather and Climate Modification*, Ed. W. N. Hess, New York, Wiley, pp. 633–86.

Index

Abbe, C., 217, 218, 219
absolute vorticity, 39, 43, 44, 47, 48, 50, 51, 212
acceleration, 4, 10, 11, 16, 17, 100, 121, 122, 123, 197; centrifugal, 179; gravitational, 206, 209; vertical, 8
adiabatic: approximation, 24–5; compression, 71; cooling, 24, 29, 185; equations, 25; expansion, 71, 181, 215; heating, 24, 84, 85, 185; lapse rate, 16; law, 16; meaning, 22; sinking, 23; temperature change, 24
adjustment procedure, 201
advection, 32; equation, 200; of absolute vorticity, 80; process, 81; terms, 16; thermal see thermal advection
advective changes, 200
ageostrophic: accelerations, 80; development process, 83, 84; flow, 81; wind, 37, 125
air motion, 3, 5, 6, 8, 14
amplitude, 146
analysis model error, 67–8, 72, 77
analytical methods of solution, 101, 195
anemograph, 2
anemometer, 130, 140
angular momentum, 179–80; balance, 124; conservation of, 179
angular velocity, 43, 100, 122

anticyclone, 19, 33, 50, 52, 75, 84, 130, 176; mid-latitude, 51–4
anticyclonic: circulation, 75; curvature, 48; development, 69, 75–85; flow, 19; vorticity, 79, 81, 84, 85, 185, 189
antitriptic winds, 123, 124
applications of analyses, 69–86
application of least-squares plane estimations, 69–74
ascending motion, 51, 54, 85
asymmetric wave structure, 191
Atkinson, B. W., 1, 8, 10, 20, 100, 101, 115–16, 121, 126–7, 153, 175, 195, 204, 212–13, 218
atmospheric: boundary layer, 140; dynamics, 2; energetics, 153–75; hydrodynamics, 118; thermodynamics, 6, 21–32, 118, 217; turbulence, 138–52
Austin, J. M., 126, 128
autocorrelation, 143–5, 150; function, 135
autocovariance, 151
available potential energy, 166–70

background field, 92–3
baroclinic: atmosphere, 118; development, 81, 120; disturbances, 72; eddies, 218; fluid, 120; instability, 127, 215, 218; motion, 181; waves, 109, 113–14